Interfaces.com

Cognitive Tools for Product Designers

Olga Werby, Ed.D.

Pipsqueak Productions, LLC
San Francisco

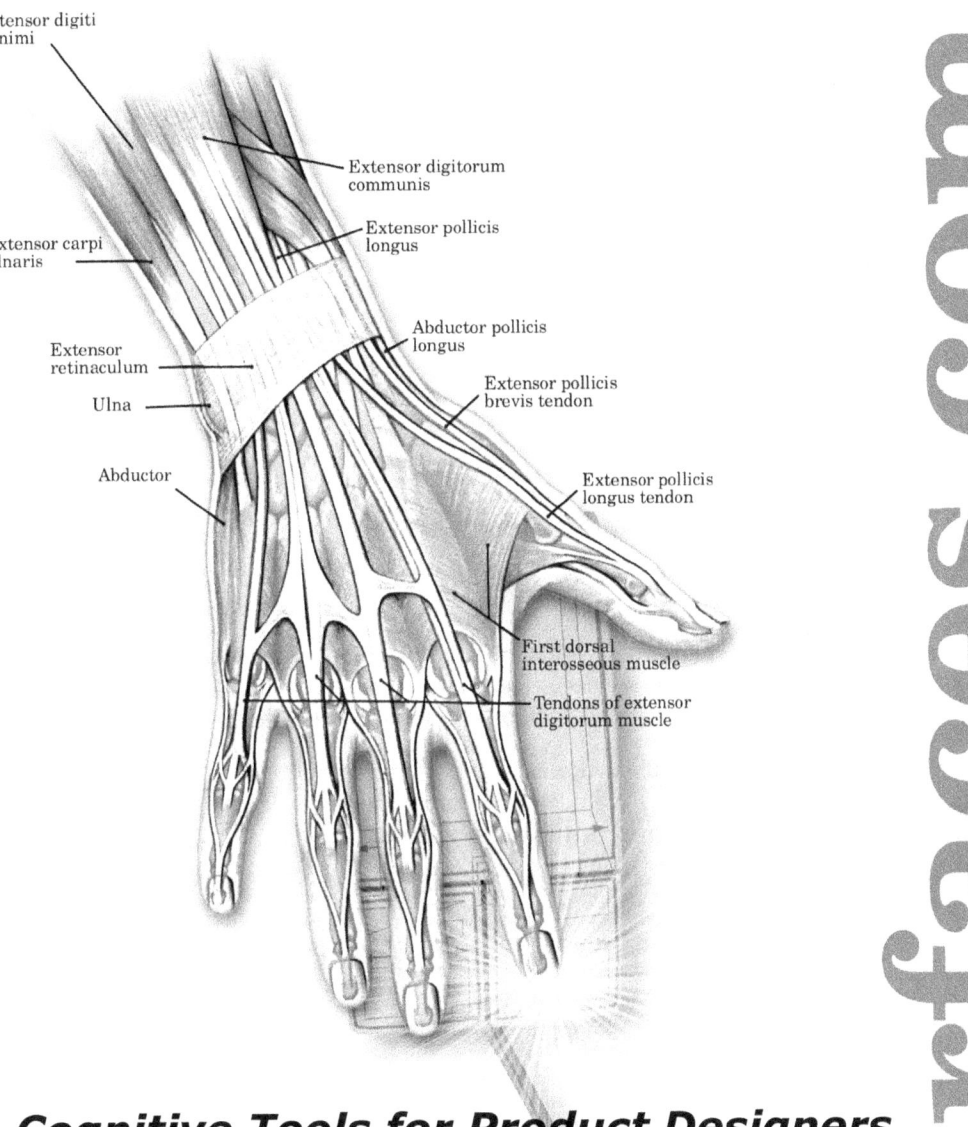

Extensor digiti
minimi

Extensor digitorum
communis

Extensor pollicis
longus

Extensor carpi
ulnaris

Extensor
retinaculum

Abductor pollicis
longus

Ulna

Extensor pollicis
brevis tendon

Abductor

Extensor pollicis
longus tendon

First dorsal
interosseous muscle

Tendons of extensor
digitorum muscle

Cognitive Tools for Product Designers

Interfaces.com

Interfaces.com: Cognitive Tools for Product Designers

by Olga Werby, Ed.D.

Werby, Olga.

Interfaces.com: Cognitive Tools for Product Designers.

Includes bibliographical references and appendix.

ISBN: 1438218036

Pipsqueak Productions, LLC is available for consulting and production work.

Pipsqueak Productions, LLC
120 El Camino Del Mar
San Francisco, CA 94121
(415) 668-4372
info@Pipsqueak.com
http://www.pipsqueak.com

http://www.interfaces.com

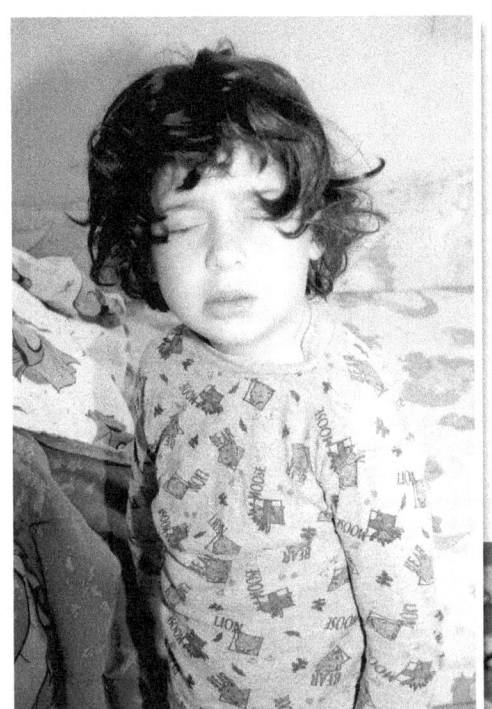

For Tim and Nick,
the morning people.

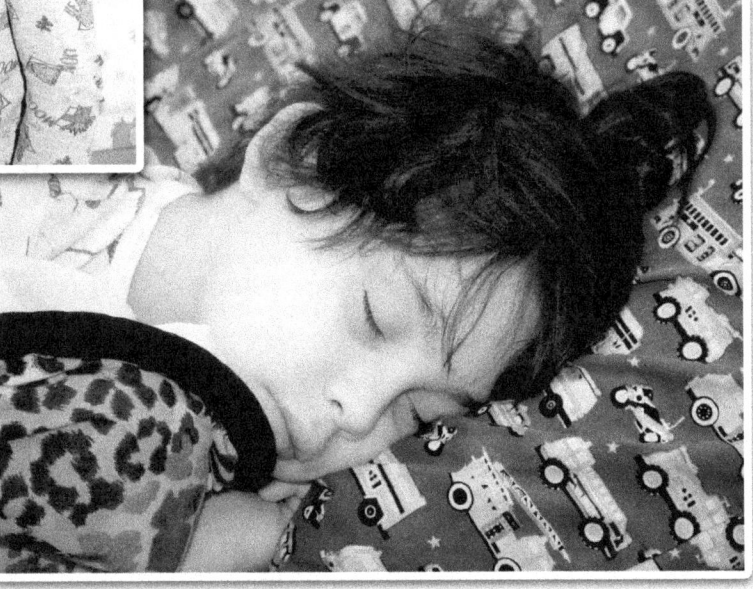

Table of Contents

Section One: Overview of Product Design

Introduction to Product Design

Virtual Product Design Overview

Product Design Approach

In The Flow

1. Introduction to Product Design

The greatest obstacle to discovery is not ignorance—
it is the illusion of knowledge.

—Daniel Boorstin

The goal of this book is to develop some intellectual tools for product designers that enable them to address common user problems. The hard work in developing products lies in understanding the cognitive needs and motivations of an audience and then meeting those needs, while shaping and fulfilling audience expectations. This book covers some recent educational theory and applies those findings to product design. In the process of examining the problems that users frequently have, the principles of effective product design are explored.

At our company, we've developed a design process that works. By working through the process and, of course, applying generous doses of talent and creativity, a designer can generate solutions that are tailored to solving the problem of the client and the user. This book discusses a lot of design theory but it does so in the service of real designers coming up with real solutions to real problems. It provides practical solutions and a methodology—a design process—that works. In the pages that follow, we'll apply this design process over and over against a wide range of product design problems. Very different issues are addressed and very different solutions are found, but the same process is used to get there.

It's difficult to talk or even think about a subject matter without appropriate vocabulary and a conceptual framework. Every profession develops a set of tools and a culture that envelops its activities, identities, artifacts, knowledge, and practice among the members of its community. This book seeks to introduce readers to the practice of product design and its mind set.

The World Around Us

As you read this, you're probably sitting down somewhere on something. If your eyesight is like mine, you're wearing glasses or contact lenses. You are probably wearing some clothing. And while you give them little attention, all of the objects that you're sitting on, wearing, leaning against, holding, and using at this moment have been designed by someone. A lot of human hours have gone into thinking, designing, engineering, producing, and introducing us to the everyday objects we come in contact with every minute of every day. We give these product creators little thought until the object we're using fails to do what we want it to do. If the chair you're sitting on suddenly collapses, its design flaws will immediately come to your attention. But rarely do we compliment the work of the designer whose product performs its duties well.

The job of a good product designer is to create products that feel so natural, so effortless to use, that we don't think about them. Such products free us to do what we want to do, free us to focus on the task at hand and not on the tools we need to accomplish it.

Initial Product Design Criteria:

- **Who is using the product?**
- **How is it used?**
- **Where is it used?**

This is a partial view of a door at the entrance to Notre Dame Cathedral in Paris, France. It's more than 20 feet tall, made of heavy wooden boards, and decorated with wrought iron detailing which also serves to hold it together and attach it to the giant hinges. When closed, the door cuts off light from the outside, dimming the interior space and highlighting the effect of the stained glass windows.

Consider a door. A door is a standard product that we encounter daily. But not all doors are the same: there are exterior doors and interior doors; doors that lock and doors that don't; doors that open and close automatically; see-through doors and solid doors; fire doors and paper screens; padded doors and painted doors; emergency exits and grand entryways; heavy doors and handicap-accessible doors; carved doors and mirrored doors; sliding doors and rotating doors; nuclear blast doors and saloon swinging doors; hidden doors and trap doors; metal doors and wood doors; air lock doors and inverse pressure doors; glass doors and darkroom doors. The design and engineering of a particular door depends on its function, its users, and the environment in which it's going to be used.

Product Design Variables:

- **Business Constraints: time, budget, talent, manpower, legal**

- **User Constraints: group size, stress, knowledge, cognitive flexibility**

- **Environmental Constraints: temporal, spatial, social, cultural, legal**

- **Engineering Constraints: materials, legal, structural, environmental**

Among other specifications, a space shuttle door needs to be airtight, needs to withstand a certain amount of vibration and flex, needs to be blasted open in case of emergency, needs to lock from the inside but not from the outside, needs to accommodate a person in a spacesuit with a set of tools, needs to have a door handle that can be operated with a space-gloved hand. These specifications are some of the design criteria for a space shuttle door design. They dictate which materials will be used for its construction, the size of the opening, its thickness, the lock mechanism, and so on. The product designer of a space shuttle door has to make thousands of decisions to insure that individuals who use it are safe.

Space Design

It is not only objects that require careful consideration. The places and spaces we live in are monuments to the design process. Architects and engineers put their stamps on many projects. City planners, building inspectors, special interest groups, parks and recreation advisors, and technical consultants are all among the hundreds of specialities that are involved in space design. But just like with any other product, this design process addresses the same set of questions: Who will be using this space? How will they be using it? What are the circumstances of use?

Consider an airport—a major hub of human and vehicular traffic. It is a collection of buildings, garages, walkways, roads, on and off ramps, storage facilities, runways, drop off and pick up zones, bus and train stops, giant luggage sorters, aircraft hangers, and people movers. Airport buildings contain police and fire stations, bathroom facilities, restaurants, storage lockers, changing rooms, waiting areas, airline counters, commerce places (regular and tariff free), first aid stations, information booths, security zones, special club rooms, check-in areas, detaining areas, lost luggage booths, baggage claim sections, multiple walkways, corridors, stairs, movers, elevators, escalators, terminals, and customs and immigration facilities. Some airports even contain museums and exhibit spaces. The structural engineering and architectural design of an airport are major undertakings.

Now consider the users of the airport. There are young children travelling alone and families travelling together, handicapped individuals and business people trying to get to their destinations. There are many languages spoken. There are experienced travellers and those that don't have a clue. Some individuals are late and some arrive incredibly early. There are sick passengers and those under the influence of mind-altering substances. There are significant security issues. And of course there are hundreds of employees, all trying to do their jobs dealing with thousands of people each day. The design of the airport has to foresee and support the needs of individuals moving through the airport to achieve their goals.

There are restricted areas, smoke-free zones, and employee only sections. It's easy to get lost, and even easier to take longer than expected to get to a particular location. Some individuals haven't slept for a very long time, and some are struggling with time differences and jet lag. Things can get incredibly tense. The design of the airport has to accommodate the emotional and cognitive states of individuals working and traveling through it.

A designer has to think beyond the needs and goals of individuals—there's the flow of people that continuously migrate through the airport. This flow changes from season to season and hour to hour. Designers have to anticipate the repercussions of these differences in flow rate and build-in supports for when this flow stops—a freak weather storm can shut down the air traffic and trap thousands of people inside the airport. What would these people need to ride out the storm?

**Munich Airport,
Internation Terminal**

Airports are built to last. Airport designers have to consider the forward evolution in aircraft development: How large will airplanes by in 20 years? How much and how fast will air traffic increase? What would be the favorite business model in the future: a set of large hubs with a web of small regional aircraft connections or a network of equal capacity nodes? It takes years and lots of capital to build airports, airports have to satisfy the business constraints and goals of their creators.

It's easy to discount these musings on airport design as too grand to apply to product design, but the principles, for an airport or a door remain the same:

- understand the needs and goals of individual users
- consider the environmental circumstances and their effects on users
- know the business criteria and motivation (e.g. time, budget, legal requirements)
- work within the structural limitations of the project (e.g. materials, zoning laws)

Cultural Differences

Outdoor fish market in Paris

An American tourist buys groceries at a supermarket in Paris. After paying the bill, she waits patiently for her purchases to be stowed away in paper bags so she can carry them home. But the clerk gets irritated with her for she is blocking the next customer and not moving out of the way. The shopping experience quickly turns sour as both the shopper and the clerk don't understand

Cultural Variables:

- **Formal Language**

- **Informal Language (e.g. idioms, slang)**

- **Rules of Etiquette**

- **Transactional Rules**

- **Business Norms**

- **Religious & Moral Differences**

- **Legal Variations**

what's stopping the other person from doing what they're supposed to do. This is not an example of a language barrier, but rather a difference in a cultural script of behavior: in America, shoppers wait for groceries to be bagged, and in France, the shoppers are supposed to bag their own purchases. When scripts collide, there is a breakdown of communication based on cultural differences and expectations. And such breakdowns are common anywhere multicultural groups of individuals intersect.

In a world where products and spaces are utilized by people of many different cultures and backgrounds, industrial design needs to understand, consider, and address these differences. The design process needs to be analyzed through the lens of cognitive science, psychology, and modern day anthropology.

Evolutionary vs. Industry-Driven Design

Consider the design of a chair.

Over eons, humans have carefully improved their implements for sitting. From logs and rocks, we moved to fine craftsmanship that doesn't only consider our ergonomic needs in its contours but also our desire for beauty. In the past, furniture was expected to last for the lifetime of its owner and beyond. Wood was chosen for its durability, color, and aesthetically pleasing grain. The height of the back rest and legs were measured and adjusted to fit the users. The width and depth of the seat as well as the curvature of the surfaces were engineered to support generations of resting bodies.

But these designs didn't spring forth overnight. Each generation of chairs had functional improvements on past versions and incorporated the cultural definition of beauty of its day. Improvements were made slowly. People had time to live with and appreciate the fine design of a chair. And as flaws in design were discovered, the next generation of chairs incorporated furniture makers' solutions to those problems. Such **evolutionary design** has the following structure:

design—test—adjust—use—get feedback from users—adjust—use—get feedback—adjust—use—get feedback—adjust—use...

For a chair, this design feedback loop spans thousands of years. Similar evolutionary design forces shaped many of our common tools: hammers, saws, files, pens, scissors, and so on.

But while evolutionary design leads to beautiful craftsmanship and positive user feedback, it's not conducive to "buy the latest upgrade" model of modern product development. People do change their furniture when they redecorate or when they feel their old stuff has become too unfashionable, but such major upheavals take place in intervals of years rather than months.

Many of the products that reach consumers today simply don't have time to evolve—the pressure to release the latest and greatest model and thus increase the company's revenues is just too great. Those companies that don't follow the rush to the market model show disappointing revenues and lose their high evaluations. And so **industry-driven design** ends up looking something like this:

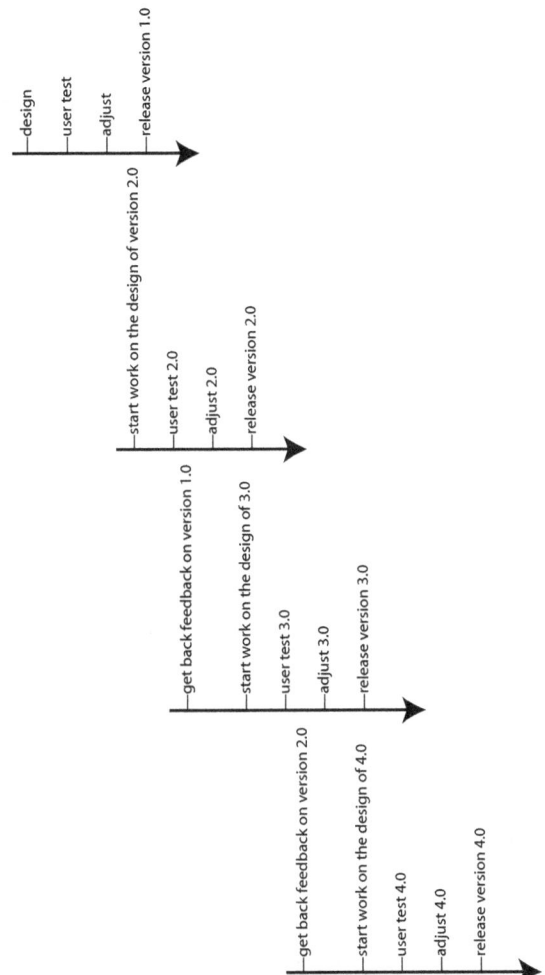

The user feedback from the first release doesn't reach the designers until they have started version 2.0 of the product—feedback is out of step with the latest version release.

Virtual Product Design

Today, product design is not limited to things we can touch and handle. Virtual products that reside only on a computer (or some version of a hand-held digital device) have the same set of constraints, variables, questions, and approaches to design. And while we don't talk about human/product interaction design, computer-based products are different. Our experience with such products is quite different from that of a door or an airport. And because our expectations of computer products are only newly emerging—our children are the first generation to grow up with the Internet and computers as part of their lives from day one—sometimes it is difficult to even tell when we experience failure. A spinning ball of death or a blank screen are all good clues to negative outcomes. But when your parent, the previous generation, can't figure out how to print out an Internet recipe, is that failure?

In the beginning, back in the dark ages of vacuum tubes and rooms of electronics, computers were hard-wired. Electrical signals were moved from one area to another through patch bays. "Programs" were changed by rearranging the patch cords. In this environment, the interface was not a high priority.

The learning curve for early computer operators was very large, and so was the error response time. As late as the early 1980s, NASA programmers frequently took 24 hours to detect a simple syntax error in the code used for Pioneer Venus data flight analysis (I know, I was there). The code would be written during the day, coded onto cards, and submitted for processing overnight. The next morning, the results were available. A program run would commonly end with: "Syntax error line 10." Repeating the process the next day would result in a program run which might say: "Syntax error line 20." Even a simple syntax error meant that the entire process would have to be started over again.

As computers moved into more general usage, there was an incentive to make them more user-friendly. In the early days of the computer operating system wars, Apple and Microsoft were allies in their battle against IBM, visualized by Apple as the monolithic big brother of the computer industry. Later, after Windows was released, the fervor with which proponents of either Mac or Windows defended their chosen platform was primarily a rallying cry around different interface standards.

As computers proliferate, it's becoming increasingly important to design simple and clear interfaces for software and web sites. Users of computer software products are far different now than they were in the 1980s. Then, software was aimed primarily at computer experts. Programmers assumed that users would become experts in order to use a particular piece of software. Most computer users only utilized a handful of computer software products to perform their work, so they could invest the time necessary to master their operation. Programmers could also assume that the users of their products shared certain cognitive abilities and expertise.

Today, users range in age from 1 to 100 and their abilities are all over the place. There is no such thing as an average computer user anymore. Users are exposed to many software products daily and, while surfing the Web, they come across dozens of different interfaces in a single session.

In this environment, "Ease of Use" has to become more than just a bonus sales point—it must become the primary consideration of anyone designing a computer-based product. If a product isn't easy to use, it will fail: most computer users won't invest the time necessary

to master its function—they'll just move on. There are plenty of other sites or products vying for their attention.

An interface is much more than a set of buttons. Navigation, while very important, is just a part of interface design. The interface is always a major part of a user's experience with a Web site or a piece of software. In some tool and application software packages, it's the users only experience with the product.

One can think of a software product as the combination of an engine, with its assortment of algorithms; the underlying content; and an interface. Traditionally, the producer was concerned primarily with the content, and the programmer was concerned primarily with the engine. The interface was often an afterthought. For tools and applications, the content is what the user creates herself. For those products, the user's only experience with the product is through its interface. The engine is in the background—and appropriately so. For products with content, like destination Web sites, the interface provides a major part of the experience that a user will have with those products.

A person uses a piece of software to achieve a goal: "I want to find and print a recipe for banana bread." The interface can either help or it can form obstacles that interfere with the realization of that goal: "Why can't I just print this out?"

The product is easier to use when its interface is designed to meet the needs of its intended audience. Designers who consider those needs produce far more effective interfaces than those who base their designs on aesthetics alone. The difference between an interface that helps or hinders a user is rarely the difference between pretty buttons and ugly ones. While well designed interfaces have an aesthetic component, interface design is more than a subset of the graphic arts.

Whose Job is It Anyway?

When I started working for a software company in the early 1980s, we didn't have an art department. Programmers created all of the graphics necessary for their products. Since there wasn't much computer graphics software for TRS-80's and Commodore 64's, this was reasonable—creating computer illustrations was very much like programming—but the games we created had a certain "pong-like" quality to them.

When in the mid 1990s my company, Pipsqueak Productions, was pitching its services for human/computer interaction design to one of the largest Web developers at the time, the president of the company didn't see a reason to hire someone other then a graphic artist to do the job. Fortunately, that company has come around in the last few years and now employs dozens of specialists in this field.

Today, a casual browse through Craigslist. com will reveal posts and requests for graphic designers, artists and illustrators, interaction specialists, human/computer interface gurus, information architects, usability experts, user experience facilitators, content creators, writers, cultural localization experts, anthropologists, sociologists, subject area specialists, cognitive scientists, programmers (with endless sub specializations), quality assurance technicians, computer scientists, product and project managers, webmasters, online learning creators, search engine optimization experts, marketing directors, and of course product designers. What do all these people do? Do they get paid the same? Who should be hired? What are the differences in all these peoples' expertise?

Job title and expertise dropping is quite common. And there's certainly an overlap between education, knowledge, experience, and work that individuals do in these different professions. Still, it's helpful to have the general sense of the differences between these job descriptions and the people qualified to do them.

The easiest way to follow the flow of work and experts is to use a concrete example, so consider an Internet-based travel agency. Let's say that after exhaustive research into the Web-based travel business, you decided to start a company selling Africa Safari Tours. From the very start, you have two very different products you're trying to develop: African Safari Tours and a Web site that will educate your customers about those tours. It's important not to confuse the two.

Product Designer

- **crystallizes ideas**

- **focuses the product**

- **articulates the business goals**

- **pins down the audience**

- **researches the market place**

- **specifies product content and components**

Who would be the first professional you might want to talk to? If you're starting a travel business, consulting with a travel agent specializing in African safari tours might be a good launching point. This is a **content area specialist**. This person might give you great ideas on what tourists want to know, how the financial arrangements work in this business, what insurance you might need to get, how much time it takes to make a sale, what you can expect to put in as your initial time and money investment, and so on. A travel specialist would also be able to give detailed information about particular locations (e.g. places to stay, people to contact, things to see) and even help plan sample itineraries for your customers.

Once you have the general sense of the travel business you're getting into and have some sample trips planned, a **conceptual designer** would help you crystallize your vision and develop a strategy of attack for your business. She will interview you and document your business goals and ideas on what you would like to do. Armed with this information, a conceptual designer will conduct research and figure out how your business plan fits within this market segment. She will then help you define your business proposition and your unique take on African safaris, as well as pin down your audience and their goals and needs in relation to your business model.

This is a critical step in any business and during any product development—someone has to actually state what the product will do, who will use it, and how the product is different from others in the same market segment. A conceptual designer generates a master plan and specifies the content you need to create for your business. With this document is hand, you can proceed to hire the rest of your team.

You might want to hire a **graphic artist** to help with your logo design. The same person can assist with gathering and editing photographs that illustrate the trips you're selling. And, eventually, a graphic artist will be responsible for the graphical look of your products. Remember, you are developing two products, safaris and the Web site to sell them. While related, they are not the same thing.

A travel **writer** can help create copy that makes your safaris sound exciting. A writer will also help with the Web copy: introduction, biographies, company background, and so on.

At this point, you're starting to build up some content and it might be a good idea to hire an **information architect**. What will he do? An information architect's job is to organize information: What kind of data do you have? What is the appropriate way to "chunk" this information? What is the natural hierarchical structure appropriate for your Web site? What labeling system should be used? What are the search terms? There are many information design tasks for a large Web site.

An **interaction design expert** will take the product design document, content, and information architecture and relate them to actual users. Taking the lead from a product designer, an interaction expert will create sample user profiles and identify their specific goals and needs as they relate to your product. Who would want to go on an African Safari? How will they approach the travel Web site selling these tours? Where will these people likely click first? Would the users of Africa Safari Tours Web site want to post photos of themselves for other visitors to check out? How would the users want to communicate with the Africa Safari Tours owner? Should the visitors have to register prior to viewing certain sections of the site? There are companies that specialize just in interaction design and nothing else—this is a deep subject area.

From the hands of the interaction designers, the product development moves towards interface design. An **interface designer** is concerned with layout and the "feel" of the site. What is the right metaphor to adopt for the Africa Safari Tours Web site? Would "Virtual Safaris" be a good approach? Should the site have light text on a dark background or dark text on a light background? Which would be easier for visitors to read given the amount of information developed by content creators for this site? Should the navigation system be on the top, left, or both? Are there any links on the footer of each page? Since the interaction designer specified a "search" feature, where should it be? Once the interface designer completely specifies the contents of each page, the size of the font, and the final navigation elements, the work is handed over to a **graphic artist**. He will create all of the visual elements of each page based on detailed specification of interaction and interface designers.

With all of the elements done and approved, the **webmaster** takes over. She is now responsible for writing code that will implement the vision of the designers that worked on the Africa Safari Tours Web site before her. A Web site is a piece of software. Its creation is not equivalent to making a paper brochure. The webmaster has to deal with browser compatibility issues, keeping the code clean and easy to update, managing effective data transfer rates, insuring against hacker attacks, building-in flexibility for future site expansion or other business needs, and search engine optimization, although there are experts that do just this part of the Web design. The job of a webmaster is very similar to a programmer. And while there are now many tools that allow non-programmers to generate Web pages, the code thus generated is often so inferior to code generated by hand that the final Web site fails to perform up to even basic expectations. For an interesting experiment, try validating the code for any random page on the Internet using the W3C code validator: **http://validator.w3.org/**

A **quality assurance** (or Q&A) specialist will take over the site when the webmaster is done. By carefully following all the links and viewing the site with different browsers and on different computers, a Q&A specialist will make sure that the site is technically correct, the links go to places they say they do, and that the site specification (as outlined by interaction, interface, and graphics designers) is followed as directed. A Q&A specialist might also be able to proofread the site and insure that there are no typos.

And finally, there is the job of a **usability expert**. Optimally, a usability expert doesn't just

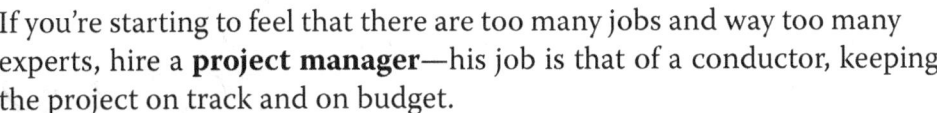

arrive at the scene at the very last minute—it's much harder (and much more expensive) to fix poor information architecture or a lousy interaction design at the end of the product development process. For each of the disciplines discussed here—information architect, interaction designer, interface designer, graphic artist, and even conceptual designer—there is a set of usability tests that can help the product remain true to its goals and the goals of its users. You will find specific questions that test these disciplines in this book.

If you're starting to feel that there are too many jobs and way too many experts, hire a **project manager**—his job is that of a conductor, keeping the project on track and on budget.

Clearly, some aspects of some of these jobs are performed by individuals who practice outside of their field of expertise. Others have been in this business so long that they can fill the shoes of many of these experts. But whoever ends up doing them, someone has to do all of those jobs, they don't get done by themselves. By specifically focusing on the different aspects of product design, you can keep tight control of the process and understand where each decision was made and why. This is good not just to satisfy your inner control freak, but it's also for the benefit of your product's success.

Additional Thoughts and Further Readings

I find that I understand and remember information better when I can apply it to something familiar. So consider this book. Like any book, this book is a product. In addition to writing, it required conceptual and interaction design, some interface work, illustrations, editing, production work, and publishing. What aspects of this book's design are covered by conceptual design? How about interface design? What professionals are needed to create this product? We will return to this over and over in the course of this book.

Entering a practice as an apprentice introduces a set of relationships between novices and old timers. For revealing insights into this social dynamic, I recommend reading "Situated Learning: Legitimate Peripheral Participation (Learning in Doing: Social, Cognitive and Computational Perspectives)" by Jean Lave and Etienne Wenger.

If you're interested in reading more about what all of these specialists do during the product design process, try Chapter 2 of "Information Architecture for the World Wide Web" by Peter Morville and Louis Rosenfeld. It provides a good description of the multiple design professions and their respective responsibilities in the product development world.

Alan Cooper wrote "The Inmates are Running the Asylum: Why High-Tech Products Drive Us Crazy and How to Restore Sanity." This 1999 book is a fun read and provides not only examples of product failures but also very good suggestions on how to conquer the design process: "A lack of design is a form of design."

"Design and Art" edited by Alex Coles is a collection of articles by "people in the business of design." The 1957 article, "Good Design: What is it for?" by George Nelson, is short and sweet. Here's a quote: "Good design, like good painting, cooking, architecture, or whatever you like, is a manifestation of the capacity of human spirit to transcend its limitation."

The Interfaith Center at the Presidio in San Francisco ran a design competition to create a space that could be used by any religious group. Some of the design criteria were compass-specific orientation, running water, sun light, non-directional seating arrangement, sitting in rows, music, silence, access to food and restrooms, and many many others. It's a fascinating design challenge. How can one space accommodate all the requirements for all religions? Solutions came from all over the world and are documented in a book: "Sacred Spaces: 2004 Interfaith Sacred Space Design Competition" and a Web site at www.interfaithdesign.org. It's an opportunity to see many multicultural design responses to a multicultural design problem.

Sacred Spaces

2004 Interfaith Sacred Space Design Competition

Edited by Donald H. Frew
Photographs by Christopher Werby

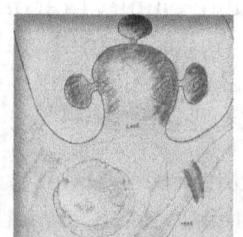

2. Virtual Product Design Overview

In most people's vocabularies, design means veneer. It's interior decorating. It's the fabric of the curtains and the sofa. But to me, nothing could be further from the meaning of design. Design is the fundamental soul of a man-made creation that ends up expressing itself in successive outer layers of the product or service.

— Steve Jobs

Historical Background

In December of 1979, Steve Jobs, then 24, was allowed to visit Xerox PARC, the pioneering research and development lab. There, legendary computer scientist Alan Kay, together with his team, had been developing nothing less than the future of computing.

In the television documentary "Triumph of the Nerds," Jobs said of the visit:

> "And they showed me really three things. But I was so blinded by the first one I didn't even really see the other two. One of the things they showed me was object orienting programming they showed me that but I didn't even see that. The other one they showed me was a networked computer system...they had over a hundred Alto computers all networked using email etc., etc., I didn't even see that. I was so blinded by the first thing they showed me which was the graphical user interface. I thought it was the best thing I'd ever seen in my life....[W]ithin you know ten minutes it was obvious to me that all computers would work like this some day."

Building on the astonishingly prescient work of Douglas Englebart from the early 1960s, Xerox's engineers had developed the fundamentals of graphical user interfaces with "WIMP" features—that's Windows, Icons, Menus, and Pointers—in 1975, four years prior to Jobs' visit to the lab. While Xerox PARC may have developed the technology, Steve Jobs and Apple Computers knew what to do with it. The WIMP-based interface became the foundation of the Lisa and, later, the Macintosh.

One advantage of WIMP-based Interfaces is that it takes a large cognitive load off the users memory by putting available operations under menus. Not only does this eliminates syntax errors but it also provides continuous visual feedback as to the current status of the system.

But windows can overlap or obscure icons or other windows. So even with a small collection of objects, a large number of different visual states may represent the same system in the same state. Try opening all the available folders on your desktop. The contents of your system haven't changed, although it's very hard now to get around and find things. If you take everything out of the cupboards of your house and spread it out all over the place, the basic contents of your house haven't changed; it's still the same stuff, now it's just messy. While open folders on the desktop maybe similar to a messy house, many novices have trouble seeing it that way. They get lost and might even think that their materials are gone.

An Interface is a Boundary

Those parts of the system that you can hit with a hammer are called hardware; those program instructions that you can only curse at are called software.

— Anonymous

What is an interface? An interface is a boundary. It's all the points of contact, physical and cognitive, between two separate systems. An expansive definition of an interface is the user's entire interaction with any designed experience. Using this definition, a book has an interface as does a supermarket, to name two examples. When viewed in this way, the essential elements which define a user's experience can be seen separately from the whole. A book's interface, for example, can be separated from the content of the book. By definition, the interface for a book is all points of contact between a book system and a human system. In the English language, pages of a book are turned from right to left. There's an index, a table of contents, and running headers and footers to help find content. Headlines, subheads, captions, pull-out quotes, and other typographic conventions are all designed to help find relevant material within a section or to otherwise enhance the reader's experience. A book has a certain weight and aspect ratio. These physical characteristics are often influenced by a designer in the choice of paper and the size of the typeface. Some books even have a built-in bookmark. And some specialized books like dictionaries sport other enhancements such as scalloped indents to aid in finding the listings beginning with a particular letter.

It is helpful to think of users as problem-solving while they use an interface. During problem solving, what a person notices is related to how she believes things work. One can think of this phenomenon in terms of the formulation: "The more you know, the more you notice; and the more you notice, the more you know." Let's take a technical support manual for a word processor as an example. The user might be trying to solve a problem, "How do I indent the first line of a paragraph?" If the user hasn't used word processing software before, he might flounder in the index searching for "paragraph," "space," "indent." A user familiar with word processing software might search for "tabs" and "rulers." The difference between these two users is their domain knowledge expertise.

The Object-Action Interface Model

There are two types of interfaces that most people come in contact during daily activities. There are GUIs—Graphical User Interfaces. And there are CUIs—Command line User Interfaces.

The Wharton Report—a 1990 usability study of GUI versus CUI interfaces—proposed seven benefits of GUIs that users experienced:

1. users worked faster
2. users completed more tasks in the same amount of time
3. users had high productivity
4. users expressed lower frustration
5. users perceived lower fatigue
6. users were more able to learn on their own and explore applications
7. users were more able to learn more capabilities of applications

The Wharton report defined a GUI as containing the following features:

1. it has a bitmapped, high-resolution computer display
2. it has a pointing device, typically a mouse
3. it promotes interface consistency between programs
4. users can see graphics and text on their screen as it looks in print
5. it follows an object-action interaction paradigm
6. it allows transfer of information between programs
7. there can be direct manipulation of on screen information and objects
8. it provides standard interface elements such as menus and dialogs
9. there is visual display of information and objects (icons and windows)
10. it provides visual feedback for user actions and tasks
11. there is visual display of user/system actions and modes (menus, palettes)
12. there is use of graphical controls (widgets) for user selection and input
13. it allows users to customize or personalize interface and interactions
14. it allows flexibility between keyboard or other input devices

GUIs can be further refined into subclasses such as geographical GUIs, desktop GUIs, etc. To be able to think about GUIs more productively, cognitive models were developed. The models give designers different ways of thinking about the GUI design problem.

The Object-Action Interface model ("OAI") requires that all the objects in an interface be defined and deconstructed down to an individual pixel (just like all the objects in the real world can be deconstructed and defined in terms of their molecular composition). And all the actions that a user can take in an OAI model start with an overall plan and then can be deconstructed all the way down to individual clicks of a mouse. OAI models work well conceptually with object-oriented programming methods and allow for easy quantification and analysis. However, I feel that starting the design process with an OAI model approach is very abstract—it removes many of the emotional and psychological aspects from interface design. Before a designer considers the color and location of a particular group of pixels on the screen, she should have thought about the audience uniqueness, the goals of the audience and the producers, how those goals can best be met through design, how users can be supported and their potential errors reduced, and so forth. From a design perspective, the OAI model seems premature. It represents the last things that should be thought about, rather than the first ones. OAI popularity stems from the fact that it can be used to evaluate interfaces that have already been built, but I feel that it offers little of value to designers facing a blank slate interaction design problem.

User-centered Design Philosophy

> *There are two major products that came out of Berkeley: LSD and UNIX. We don't believe this to be a coincidence.*
>
> —Jeremy S. Anderson

User-centered design philosophy puts the users' needs at the center of the design process. It states that design is iterative in nature and users should be considered at multiple stages of product development. Designers are encouraged to try to think like users when coming up with design solutions. These solutions are then exposed to real users. Testing reveals some of the design flaws. Designers then are supposed to correct them and iteratively go through this process again and again.

But the user-centered design philosophy doesn't give designers much help in trying to actually understand why a particular design fails. User-centered design philosophy doesn't provide the designer with a set of tools to attack user problems once they are discovered. And, while testing is always desirable, iterative testing and development can be quite expensive, and is often a luxury that designers don't have in the real world.

User-centered design isn't so much a theory as a goal. It is certainly desirable to put the user at the center of product design. How this is done, however, is the real question.

An Interface is Where a User Solves Problems

At my company, we developed the "Interface as a Problem Solving Medium" model, similar to user-centered design but with an emphasis on cognitive science. It starts by focusing on the end user. While I'm describing GUI interfaces, this model also applies to CUIs, hardware devices, product design, or to supermarkets and playground design for that matter. Nothing limits this model to computer-based interaction.

This model assumes that when people sit down to do a task on a computer, they are trying to solve problems with the interface. Perhaps they're searching for information or they are creating something—either way, they are using tools to achieve a desired goal. Consider a word processor. Once it's loaded, the font is chosen, the margins and tabs are set, and the user starts to actually write. At that point, the writer stops interacting with the interface and instead focuses on using the tool. But as soon as the writer needs to do something else—erase a word or move a sentence—the interface issues again come to the fore. And, especially for novice users, those tasks are all about problem solving.

The problem solving model focuses on the people who intend to use the interface—the end users—and I feel it is therefore conducive to helping designers produce a more satisfying and user-friendly interface.

Whichever model is used by a product designer, having a particular perspective gives the designer a handle on how to approach this complex problem. So if an interface is a medium in which people solve problems, an interface designer should know how people go about solving problems, the limitations on their abilities, and what constructive help can be given to them through the design of the interface.

Additional Thoughts and Further Readings

This chapter focused on computer-based products. So consider a virtual book. What are the interface features of such a product? How would they be different from a "real-world" book? Can we even talk about a paper-back novel as having an interface? Does it help with the design?

There are hundreds of books written about human/computer interface design, most just in the last two decades. Theo Mandel's "Elements of User Interface Design," published in 1997, and Ben Shneiderman's "Designing the User Interface," reprinted in 1998, are standard introductory textbooks in this field.

3. Product Design Approach

Wisdom is knowing what to do next; skill is knowing how to do it, and virtue is doing it.

— David Starr Jordan

How does one approach product design? What needs to be happen first? How does the process of design start? What are the business goals for the product? What do designers hope to achieve by developing this project? Who is the intended audience for this experience? Are the goals of the project well defined? The very first step is listening.

There are hundreds of questions and many times these questions not only don't have an immediate answer, they are not even asked. It's the job of the product designer to carefully guide the project and its developer through the investigation and to develop guidelines for the whole project before the design process really begins. If the goal of the project keeps changing over the life of development and execution, it will be almost impossible to come up with a carefully thought-out and well-aimed design.

To get a handle on product design, it is helpful to break the task down into three components: Conceptual Design, Interaction Design, and Interface Design.

Conceptual design answers the question "What does this product do?"

Interaction design deals with "How does the product do it?"

And **interface design** focuses on "How does the product look and feel?"

Early in a product's creation, conceptual design is critical. Later, as the feature set is defined, interaction design and interface design predominate.

Conceptual Design

Conceptual design focuses on what a product does. The start of the process is identifying the audience and then visualizing what that audience wishes from a particular product. Then we can identify the goals and expectations of each audience type under the different circumstances of use. We can then seek ways of satisfying those user goals and shaping those expectations. Rather than driving this process by what can be done technically, it's best to adapt an "audience experience" approach to determining a product's feature set.

Take as an example the creation of a Bank Automated Teller Machine (ATM). There are three focal points to this device: the touch screen which allows input and provides visual feedback; the keypad that allows the user to enter information; and the physical input devices which handle the manipulation of the bank card, the acceptance of a deposit, and the delivery of the receipt and currency.

Conceptually, an ATM needs to allow an ordinary person to do some minimal banking functions: deposit and withdraw money from multiple accounts, transfer funds between accounts, and receive information on check balances and recent transactions. Security is critical. ATM security can be divided into bank security functions and the customer's physical security. During the conceptual phase, the designer needs to acknowledge the vulnerability of the bank customer standing outside with her account and money exposed

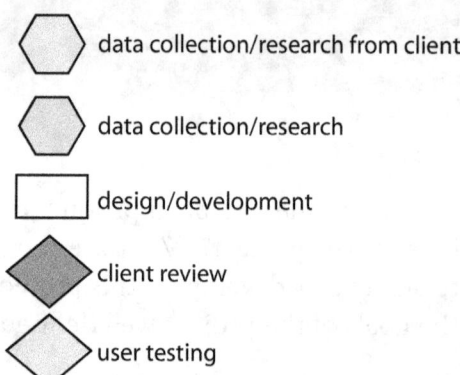

data collection/research from client

data collection/research

design/development

client review

user testing

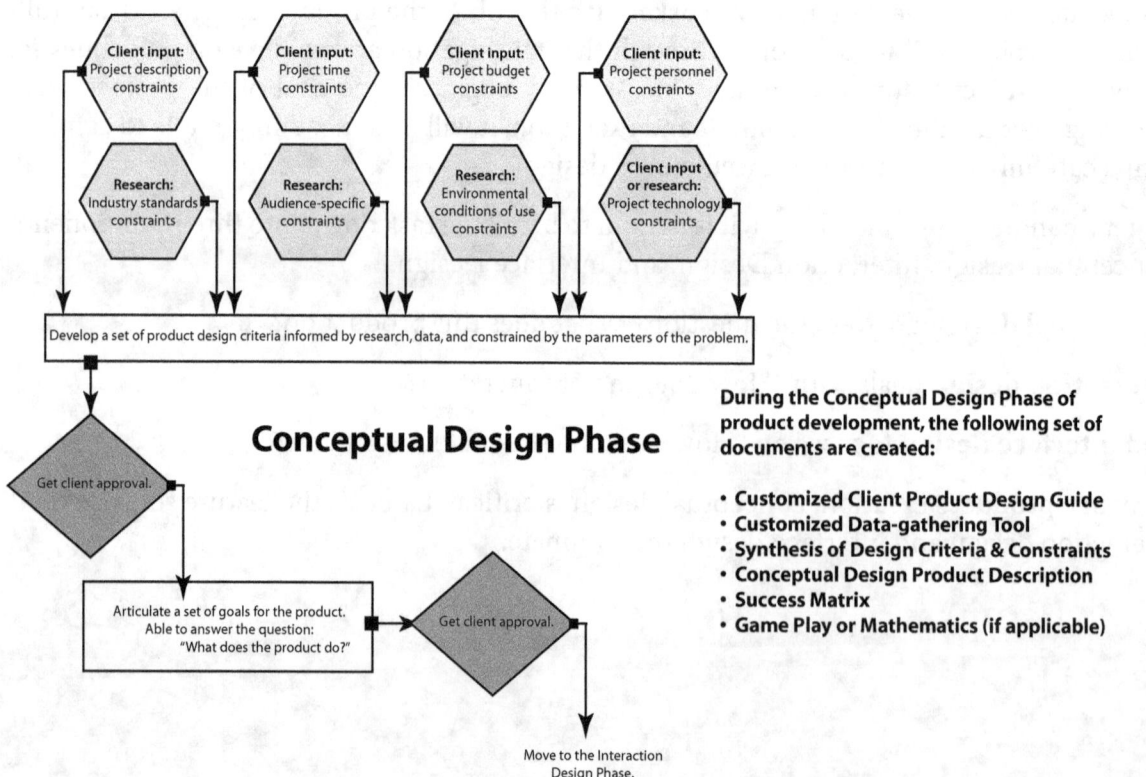

Conceptual Design Phase

During the Conceptual Design Phase of product development, the following set of documents are created:

- **Customized Client Product Design Guide**
- **Customized Data-gathering Tool**
- **Synthesis of Design Criteria & Constraints**
- **Conceptual Design Product Description**
- **Success Matrix**
- **Game Play or Mathematics (if applicable)**

and help prevent a physical attack. Bank security, during this design phase, deals with protecting bank assets.

The conceptual design specification of an ATM would list all of the desirable functions that the machine would do to meet the needs and goals of its users and of the bank. It would also make certain cost/benefit decisions. For example, on the customer security issue, it may be desirable but impracticable to provide a security guard at each ATM location.

Interaction Design

Interaction design focuses on how a product behaves. It touches all the points of contact between a product and the user. As such, it unlocks the functionality of a product. Information architecture is part of interaction design. For example, during the interaction design period of a Web site, the designer might determine how the Web pages organize the content.

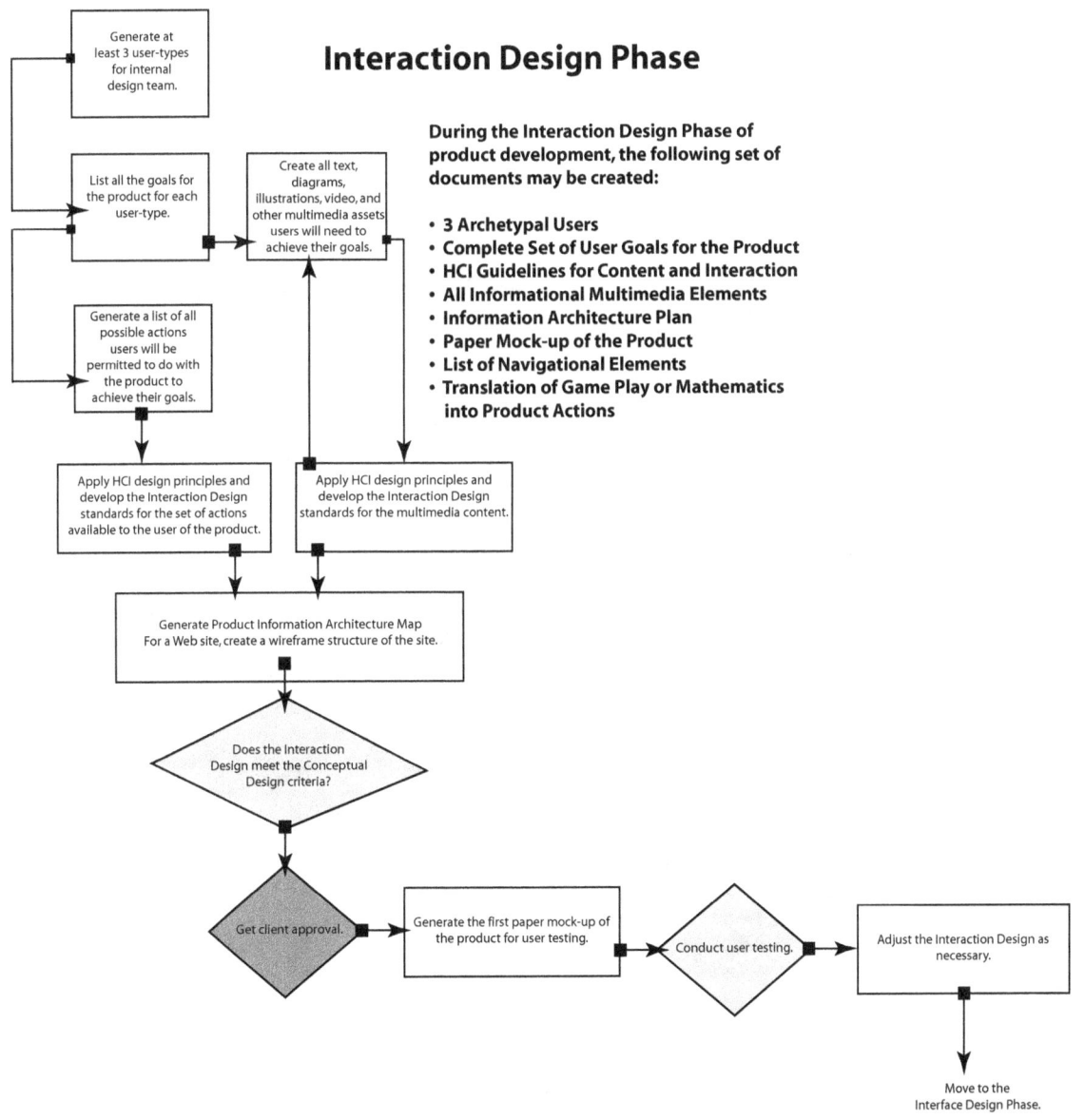

In the ATM example, once the list of features has been drawn up, the team working on the interaction design of the ATM will figure out how to best implement those features. During this stage, it would be important to figure out things like how big to make the "hot spots" on the touch screen, how the screen should be angled to reduce glare, which information should be entered on the keypad and which via a touch screen, how many levels deep to make hierarchy of information presented to the user—too deep, and most people won't be able to navigate it.

Responding to the ATM customer security issue raised during the conceptual design state, for example, the interaction designer may specify site lighting, television cameras, screens viewable from only a narrow angle, privacy panels, reflective surfaces to allow users to see what's going on behind them, and so forth.

The list of all interaction design criteria is very long even for a relatively simple product. How these interaction design criteria get implemented is the job of interface design.

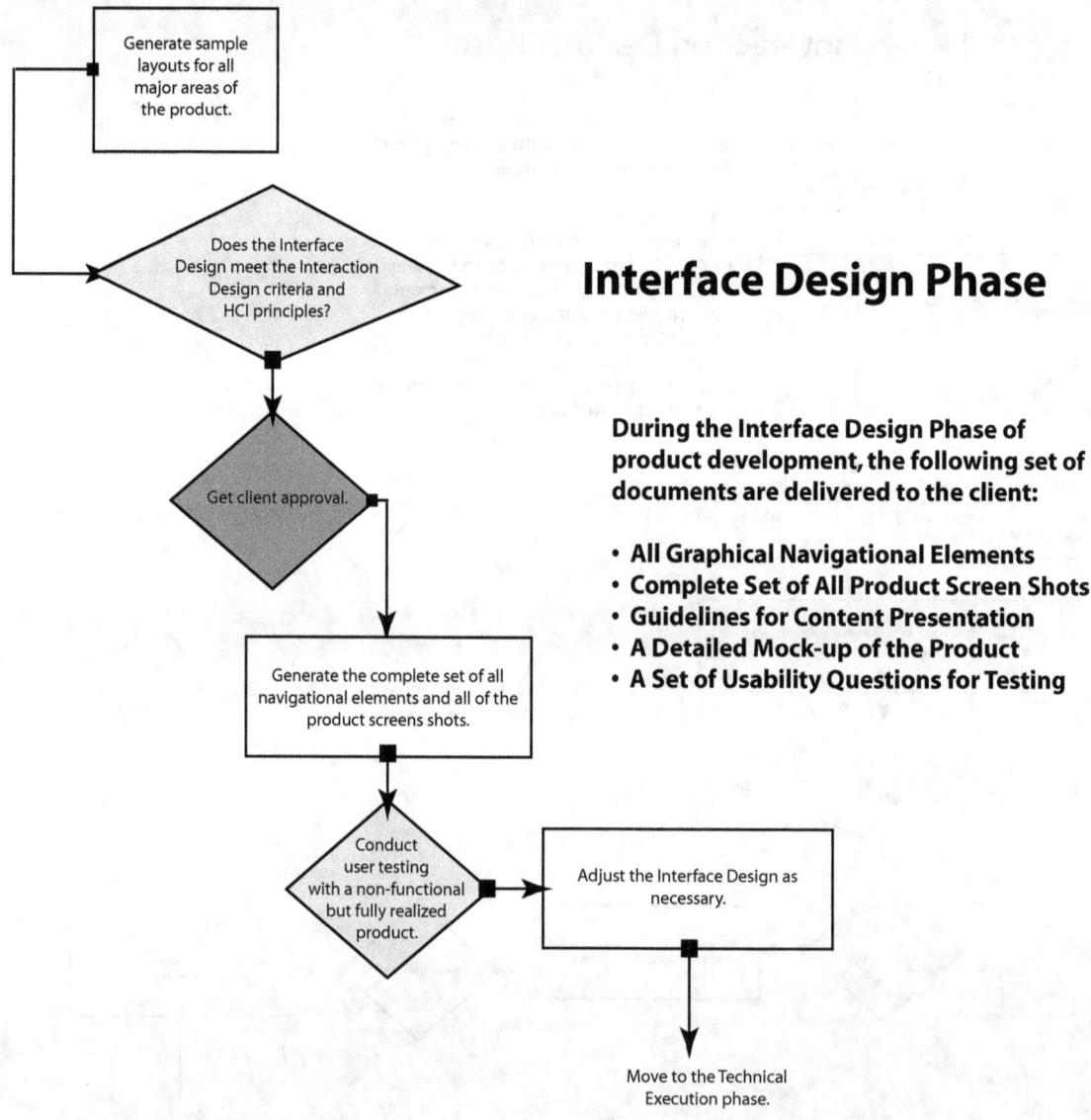

Interface Design Phase

During the Interface Design Phase of product development, the following set of documents are delivered to the client:

- **All Graphical Navigational Elements**
- **Complete Set of All Product Screen Shots**
- **Guidelines for Content Presentation**
- **A Detailed Mock-up of the Product**
- **A Set of Usability Questions for Testing**

Interface Design

Interface design focuses on how the product looks and feels. Often the problems of look and feel are mistakenly thought to be the entire design problem. But look and feel is only successful if the other design issues have been carefully crafted first.

For the Bank ATM, interface design includes the materials out of which to make the keypad, the way the information is presented on the screen, and the choice of user feedback for each action the user makes. Recently, Wells Fargo replaced its ATMs with a newer touch screen model. While similar to the old, the new ATM model differentiated between the beep sounds the keypad made and the beep sounds of the touch screen. So if a person wanted to withdraw a standard amount, the beeps revealed only that she was pressing the touch screen (one beep per standard amount). But if the user wanted something different, she would have to use the keypad, and the sound of the beeps revealed what she was doing—five beeps meant that the withdrawal was $100 or more. The physical security of the customer is potentially jeopardized by the sounds of the beeps. This is an example of a flaw in the interface design stage.

Often, product designers juggle all three aspects of the design at the same time—you can't produce a good interface design if the interaction design is lousy; and it is very difficult to create a quality interaction design without having a good understanding of the function and goal of the product. Separating these steps into consecutive design stages makes for better products and a more efficient development process.

Product Design

So what do you do first? How do you start on the project? If you are working with a client, you have to start by listening. What does a client really want? What are they hoping to achieve by developing this project? Who is their intended

Conceptual Design: What does this product do?

Interaction Design: How does the product do it?

Interface Design: How does the product look and feel?

audience? Are the goals of their project well defined? There are hundreds of questions and, in our experience, many times these questions not only don't have an immediate answer, they haven't even been asked yet.

When you are both the client and the designer, the problems don't go away. You'll have to pose and answer the same tough questions. The scope of the project needs to be

established, and a set of features needs to be chosen and reduced from a million great ideas. While you have creative freedom, the paradox is that sometimes the restrictions imposed by a client act as a spur to creativity. When working on your own projects, there's a strong temptation to pick the first workable solution rather than continue a search for the best one. Resist this temptation. Don't fall in love with your ideas.

Little Birds Day Care Logo Design

What if you're charged with creating a logo for "Little Birds Day Care?" How can the above process apply to this concrete situation? Let's start with conceptual design. This imaginary day-care provides services for parents and their children between two and five years of age (at which point the children start school). While we can assume that parents are pretty good readers, their children probably haven't yet mastered this skill. But it would be good if both the parents and their children were very good at identifying the logo of the day-care: if lost in a museum, for example, a child should be able to identify members of his group by the logos on their buttons or stickers. If the logo has a picture component, this task would be much easier for a non-reader. Thus the first criteria is to develop a word mark and a graphical mark that would make up the complete logo of "Little Birds Day Care."

Conceptual design for "Little Birds Day Care" logo might also specify that the birds in question are swans—the owner of the company likes swans. Since this company deals with children, the logo mark should also feature little swans, cygnets, together with an adult swan. The adult swan should look "caring and nurturing."

Interaction design can further zero-in on the logo mark design. There should be three baby swans—more is too much, less and it doesn't look like day-care. Interaction design can also talk about the tone of the desired illustration: one might be "magical, Victorian, detailed", another might be "loose, suggestive of different cultures." It might also include technical specifications like the palette, resolution independence, and the ability to be faxed.

Interface design would specify the look and feel. One approach may be a pen and ink drawing. Another might be black brush strokes.

After it's designed, the logo needs to be executed—that's production, in this case graphic arts. Here is an example of two "Little Birds Day Care" logos, both of which satisfy the

above mentioned conceptual and interaction designs, but diverge in the graphic execution. A graphic designer has a lot of freedom even after conceptual, interaction, and interface designs criteria have been specified.

While graphic design doesn't equal conceptual, interaction, and interface design, there is clearly a continuum of thought from the initial idea right through to the final execution. There are people who can do many parts of this process. But it's important to recognize the wide variety of the different hats they wear. Each requires specific expertise to do well.

Additional Thoughts and Further Readings

To arrive at the simple is difficult.

—Rashid Elisha

This chapter broke down the design process into three distinct steps: conceptual design, interaction design, and interface design. This process is described in detail in Alan Cooper's "The Inmates are Running the Asylum: Why High-Tech Products Drive Us Crazy and How to Restore Sanity." Alan runs his own design firm that specializes exclusively on interaction design: www.cooper.com

Conceptual and interaction designs can and should be considered separately from interface and graphic designs. Intended as a technical demonstration of semantic markup separated from CSS presentation, www.csszengarden.com also is a wonderful demonstration of how the look and feel can be completely separated from the content of the Web site.

4. In The Flow

There is no reason for any individual to have a computer in their home.

—Ken Olson (President of Digital Equipment Corporation) at the Convention of the World Future Society in Boston in 1977

Computers in the future may have only 1,000 vacuum tubes and perhaps only weigh 1.5 tons.

—Popular Mechanics, 1949

Optimal Experience

What does it take to have a good time? Ask a hundred people, get a hundred answers. A sizable percentage of the sample group might mentioned sex. Others might list their favorite activities: reading, talking with friends, working on a hobby project. Playing video games would be on our sons' lists. What do these activities have in common? How can we capture the essence of a "good time" and duplicate it in our product design?

Let's take sex and reading. How are they the same? When both are enjoyable, they cause us to lose track of time. We get totally "into" it. That's certainly true of video games as well. And I feel that way about painting, my favorite hobby. To consume our total attention, the activity can't be boring. Our minds can't wander to extraneous subjects—if you're picking your wardrobe for tomorrow during sex, you're not really enjoying it. The activity has to be **engrossing**.

Total attention is not enough. You can be at the dentist getting drilled and unable to take your mind off your current predicament. Pain is a wonderful focuser, but it is usually not associated with "good time." So the activity has to be something we **enjoy**.

Optimal Experience:

• **Enjoyable—emotional variable**

• **Engrossing—attentional variable**

• **Environmentally Appropriate—situational variable**

• **Cognitively Exhilarating—skill variable**

I love Sudoku. I've spent hours solving Sudoku puzzles, happily drinking coffee on a lazy weekend morning. But I don't like all of the puzzles. Easy ones are just too easy, there's just no challenge to doing them. The "impossible" ones can be okay, but they take so long that I start to feel guilty wasting my time when so many other things need to get done (Sudoku puzzles have to be done in one sitting). So the impossible ones don't qualify for a "good time," but for completely different reason than the easy ones—they don't fit the parameters I set for playing this game. So **environmental conditions** matter, once again.

My sons are avid chess players. It's one of their optimal experiences. They are very good, too. But they are not interested in playing "newbies"—what's the fun in the game that's over so quickly? And while playing highly-rated players is a lot more interesting, if there's no chance of winning the game, then the pleasure is greatly diminished. Chess players have to be equally matched to experience optimal enjoyment.

This is the Goldilocks' problem. The activity can't be too easy or too hard. It has to fit the skill set of the individual just right. It has to be **cognitively exhilarating**. If it holds too little challenge, we get bored, our minds start to wonder to other concerns and we lose the connection to the experience. If it is too difficult, then we get stressed out—there's no pathway to success.

Mihaly Csikszentmihalyi calls the experiences that satisfy the above criteria **flow**. And flow is what we want to build into our products. If the products we design cause flow and produce optimal experiences for their users, then those products have a high chance of success in the marketplace.

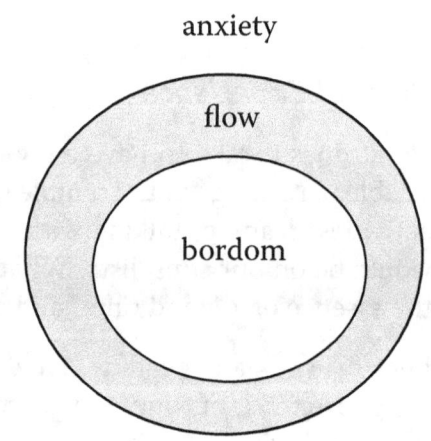

anxiety

flow

bordom

Games

What makes a game fun? What keeps us playing for hours into the night? Why do some games feel so addictive? If you are a game designer, you want to know.

When you design a functional tool or an informational web site, the main goal is to make it easy to use. This means understanding the cognitive limitations of the audience, their goals, and their expectations. If something is too hard, cognitive scaffoldings are set up to support the users in their endeavors. But for games, the reverse it true. Part of the fun of playing a game is that it forces the user to strain their cognitive abilities to the limit. So while a complicated labyrinth would be inappropriate for CNN's information architecture, it might make an adventure game more exciting by stressing the gamer's attention controls and working memory. If an investment site jumps out at a user and makes them anxious, that's not good design. But when a thriller does the same, that's a good game. When a shopping site defies expectations, the result is poor sales. But when a game does it, it might become a best seller. The trick is to stay in the flow.

So part of what makes a game fun to play is how cognitively challenging it is. Clearly, a game could be made too difficult, but some individuals thrive on that complexity. As a simple example think of the London Times Crossword Puzzle. It's a fiendishly hard entertainment. And yet millions look forward to it and compete for the fastest solution times.

Emotional Intelligence of a Web Product

The authority of those who teach is often an obstacle to those who want to learn.

—Marcus Tullius Cicero (106-43 BC)

We've all met people who are trying to be nice, but aren't. Something about them comes off as unpleasant—and it has nothing to do with the content of their conversation or the way they look or dress. Psychologists sometimes refer to such people as socially challenged, or lacking emotional intelligence, or emotional quotient ("EQ").

Just like content, an interface has an emotional intelligence. A Web site has an emotional impact on its audience. When they leave, most people's primary impression won't be a memory of the informational content. Rather, they'll retain a vague feeling of "good" or "bad" evoked by their experience.

If you're designing a Web site, you should consider the emotional intelligence of your site. Is your user a happy user? Do people leave your site frustrated or satisfied? Is your home page overwhelming, confusing, or even worse, is it scary? Are you intimidating your users?

Emotion is a strong memory trigger—emotionally charged memories are easier to recall. So what can Web site designers do to make a site memorable? How can we capitalize on emotional intelligence? And what should we avoid?

Some things are obvious:

- Frustration on the Web builds negative perceptions. If anything goes wrong, even with the user's own connection, it's often perceived as the Web site's fault. There's nothing the site designers can do about some of this, but designers can work to reduce frustration where possible: producing efficient low bandwidth pages that load quickly, creating links that work, avoiding technology that delivers zing but can crash the user's computer—these are a few basic principles of an emotionally savvy Web design.

- Users are suspicious any time they are asked to give out personal information. They'll scrutinize each request to evaluate if the site really needs the information and to see how they can be hurt by divulging it. Any suspicion that they're being conned and the users' trust in a site will be gone.

- Have you ever written a letter and had it completely ignored? How do you feel about a person who doesn't answer your correspondence? Now hold on to that feeling. If your customer service department doesn't answer your visitors' email inquiries, they will feel about you the same way.

- Be polite. There was an interesting experiment where two groups of beta testers were asked to evaluate a program. The computer product each group saw was identical except that the dialog boxes and error messages for one were particularly polite—saying "please" and "thank you." The beta testers of the polite product not only liked it more but also reported it as both more functional and stable than the other version. So be polite. A page on the Web is seen as more personal than a printed page of a brochure.

- And finally, you need to know your audience—if the product is designed for a professional audience of physicians, don't talk to them as if they are children. Similarly, it's important to talk to patients in a non-condescending manner that uses language within their level of expertise. Finding the right tone is one of the important jobs of both writers and designers of Web sites.

An emotionally intelligent product will:

- be well equipped to deal with its audiences' frustrations and fears
- set the right mood
- use the entertainment value of the product to motivate the user of the product in a difficult situation—i.e. motivate the user to do hard work
- change indifference or casual interest in a product into motivation
- have a manual or online help that uses humor, keeps the audience on the same social level as the product's producers, without being condescending

- keep control over its user but create a feeling of personal choice

- create an emotional high and "ah ha" moments to make its users' experiences more memorable

We just bought a new washing machine from Maytag and are in the process of registering the purchase with the company. I got about halfway through the registration and have abandoned the process. As part of the registration, the company wanted to know my birth date, how many children I have and their ages, my income level, if I own my house, the appliance I'm likely to buy in the next year, my profession, and my marital status! I found this so outrageous, that the good happy feelings I had for my washing machine are now almost gone.

A Web page is viewed as a strong reflection of the company it represents. Customers don't view a poorly designed or nonfunctional page simply as evidence that the company doesn't know how to do a Web site. Rather, they directly translate their negative experience with a Web site into a negative view of how the company conducts business generally. What a customer takes away from a technologically troubled site may not be rational, fair, or true, but it's a perception they'll act upon nonetheless.

Design doesn't end when the last bolt in screwed into place, there are all those supporting materials that can enhance or take away from the experience with the product. Thus one of the nominated words of the year for 2007 was "wrap rage"—"anger brought on by the frustration of trying to open a factory-sealed purchase" (American Dialect Society 2007 Words of the Year Nominations).

Transparency and Thickness of a Web Site

In thinking about how Web sites are perceived by their users, I came up with two underlying properties which, once understood, can be manipulated to strongly affect the **emotional IQ** of a site. The first of these properties I've labelled as a Web site's "**transparency**" and the second I've labelled as its "**thickness**." Both refer to the users' feelings in their own relation to the site.

Transparency, and its antonym **opacity**, form two opposite poles of one continuum. If a Web site is opaque, then the user feels that her identity is shielded from both the site creators and other users. She is anonymous. If a Web site is transparent, by contrast, the user feels known to both the site creators and the other users. A feeling of transparency can be desirable in a community site, for example, while a pornographic site might be better served by wrapping its users in a feeling of anonymity/opacity. Most sites are somewhere in between.

www.Amazon.com, for example, stores information about its users and tries to entice them with other books they feel are related to the ones previously purchased. But although a user is tracked through the books that they browse, the site is designed so that the surfer is uninvolved with that tracking. If Amazon.com becomes too transparent, users may avoid purchasing or browsing certain books due to embarrassment. Obviously this is true with racy titles, but other titles might also be impacted by Amazon's choice of where to sit on the transparency/opacity continuum. For example, Amazon's "recommend a book based on previous purchases" feature sometimes fails. Some customers will decide not to buy a book because they don't feel the book matches their self image and they don't want it factored into those recommendations.

The TiVo television recorder device tries to predict programs that the user might like based upon programs previously requested by the user. A famous "Seinfeld" television episode used this "recommend a program" feature as a storyline where the character was asking TiVo to record a host of "manly" programs because he was convinced that his TiVo thought he was gay.

Similarly **thickness**, and its antonym **thinness**, also define a spectrum along which a Web site can reside, hopefully by conscious choice. A site is thin if the user feels that the creators of the site are very accessible to him. A site is thick if the user feels far removed from any humans on the other side of the Internet connection. A good customer service site will strive for a thin interface, to give the customer the sense that humans are highly accessible. While surfing a Web site with a thin interface, the user doesn't feel as if barriers have been erected to

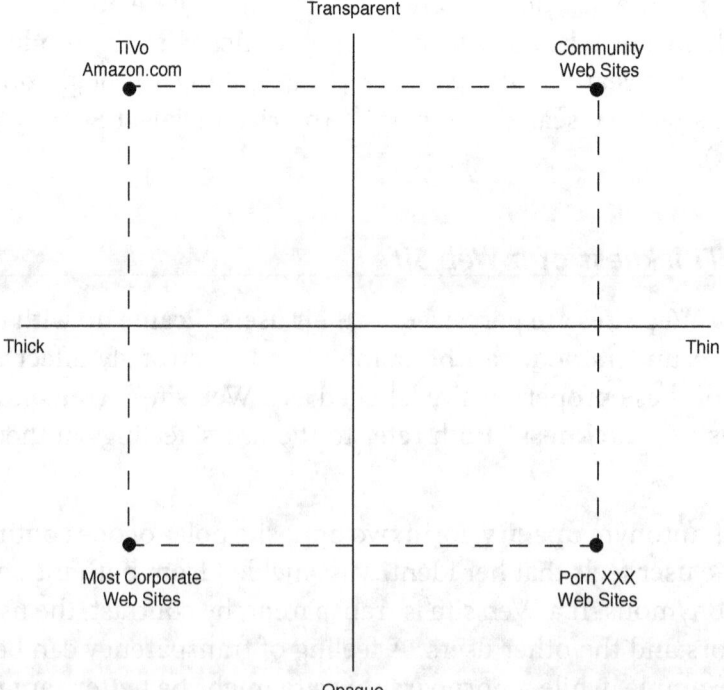

keep her at a distance. By contrast, a lot of large corporate sites feel thick to outside visitors. Corporate officers often keep their email addresses secret because they don't want to receive personal email from strangers or, perhaps, their customers. Thick interfaces often feel more formal than thin ones.

Transparency and thickness are just some of the qualities that go into defining the emotional IQ of a product, but I've found them to be useful concepts, and we consider them carefully during any Web design project. The diagram below shows how these concepts can be expressed graphically for the design phase and for the pre-launch questions asked of customers.

The emotional intelligence of a Web site gets easily overlooked in a board meeting, but it doesn't get overlooked by the site's audience. The most important step in raising your site's emotional IQ is figuring out who are the people who will be visiting and using your site, what they want, and how you can give it to them.

Scaffolding

So it is that we all live with minds wired to excel in one area and crash in another. Hopefully, we discover and engage in good matches between our kind of mind and our pursuits in life.

—Mel Levine

A situation or a task is made more difficult to a particular user when it is presented outside of their **cognitive comfort zone**. Any user is outside of his cognitive comfort zone when he is asked to do things he doesn't know or understand—outside of his area of expertise. For example, nonnative English speakers might need a dictionary—an **informational scaffold**—to understand a cooking recipe even when they have a professional culinary background. But a novice chef might need help understanding a recipe even without a language barrier. A user is outside of her cognitive comfort zone when the memory demands of a task overwhelm her memory capacity. For most people, multiplying triple digit numbers in their head is a nearly impossible task. And a person who is not an "audio learner" drowns in a sea of words and sounds during a lecture—he is outside of his cognitive comfort zone in that situation.

Scaffolds are cognitive support structures that aid users in accomplishing their goals in difficult situations. One example frequently used in interaction design is preparing an analogy to something the user already knows well, or perhaps providing a metaphor for the new experience based upon the individual's prior experience. The files and folders of graphical operating systems are an attempt to analogize to an office environment, for example. But a metaphor or an analogy has to be within the reach of the user—analogizing to an office environment for a second grader might not generate a lot of helpful supports.

Since the cognitive comfort zone is different for different users or groups of users, the scaffolds have to be closely targeted to meet the cognitive needs of the product's audience.

Scaffolds are employed by designers to support and enhance the user's ability to perform certain tasks. For example, Microsoft Word comes with a slew of sample layouts that novices can use as a starting point for their own page designs. As their skills grow, the samples become less useful because it becomes just as easy to create new designs from scratch. The goal for a well-designed scaffold is to fade into the background as the need diminishes: as novices turn to experts, supports should gracefully recede.

Scaffolds are important, for they give a novice a chance to actually produce a product or achieve a goal that would otherwise not be possible. The sense of accomplishment felt by a user during the early stages of learning a piece of software can help them in their effort to master it.

Scaffolds are not only an important consideration during the conceptual design stage, but they are also useful tools during the interaction and interface design stages. As a matter of conceptual design, for example, you might try to avoid requiring the user to multiply three digit numbers in her head. But if that is required, the interaction with a product can be designed to break the multiplication activity into more manageable steps. In short, the design can be created to assist the memory from becoming overloaded. Once cognitive needs are identified, designers can search for appropriate solutions.

Additional Thoughts and Further Readings

Mihaly Csikszentmihalyi wrote "Flow: The Psychology of Optimal Experience" in 1990 (later published by Harper Perennial in 1991). It's an interesting read, both fascinating and frustrating at the same time. But the main idea, the idea of flow experience, is a powerful one and should be considered during product design.

If you're a Sudoku fan or just like to see what it is, try these Web sites:

> www.number-logic.com
>
> www.sudoku.com
>
> www.websudoku.com
>
> www.dailysudoku.com

Section Two: Users

What Users Bring to Usability

Introduction to Memory

Working Memory

Long Term Memory Filing and Manipulation

Background Knowledge

How We think About Probelms

Personality

Perception

Language and Linguistic Processing

Attention & Impulsivity Controls

5. What Users Bring to Usability

The pure and simple truth is rarely pure and never simple.

—Oscar Wilde

We all arrive on the scene with different baggage—our experiences, education, perception, memory, and so on are unique to each of us. No two individuals interpret an experience in exactly the same way. While this sounds daunting, we shouldn't give up on design altogether. We all have some things in common. To examine differences and commonalities, we need to examine how people think. Cognitive science and psychology provide some helpful answers.

Everyone brings their own baggage.

Cognitive Wheel

Product Design Variables:

1. **Environment**

2. **Social Knowledge/Culture**

3. **Memory**

4. **Domain Knowledge**

5. **Organization of Knowledge**

6. **Perception**

7. **Attention Controls**

8. **Language Skills**

9. **Personality**

Human cognition is a famously complex process. It's hard to explain what makes an individual good at something; what makes her a good learner, or a good thinker. Different scientists studying cognition have developed different taxonomies depending on their professional goals and the aspect of cognition that interested them. In my work, I focus on using cognitive science to engineer educational solutions, to design better products, and to improve human/computer interactions. Thus, I developed what I call the Cognitive Wheel as a way to think about cognitive attributes in the context of product design.

I've summarized research in education and cognitive science into a set of variables I find most useful to consider during product design. I broadly divide cognitive attributes into the following set.

Background Knowledge
- formal and informal background knowledge of a particular domain
- cultural and social norms
- meta knowledge

Perception
- perceptual style
- perceptual processing

Personality
- anxious
- collaborative
- competitive
- computer affinity
- expressive vs. reserved
- observant vs. introspective
- schedulers vs. probers

Memory
- short term memory
- working memory
- long term memory

Attention Controls
- mental energy controls
- input controls
- output controls

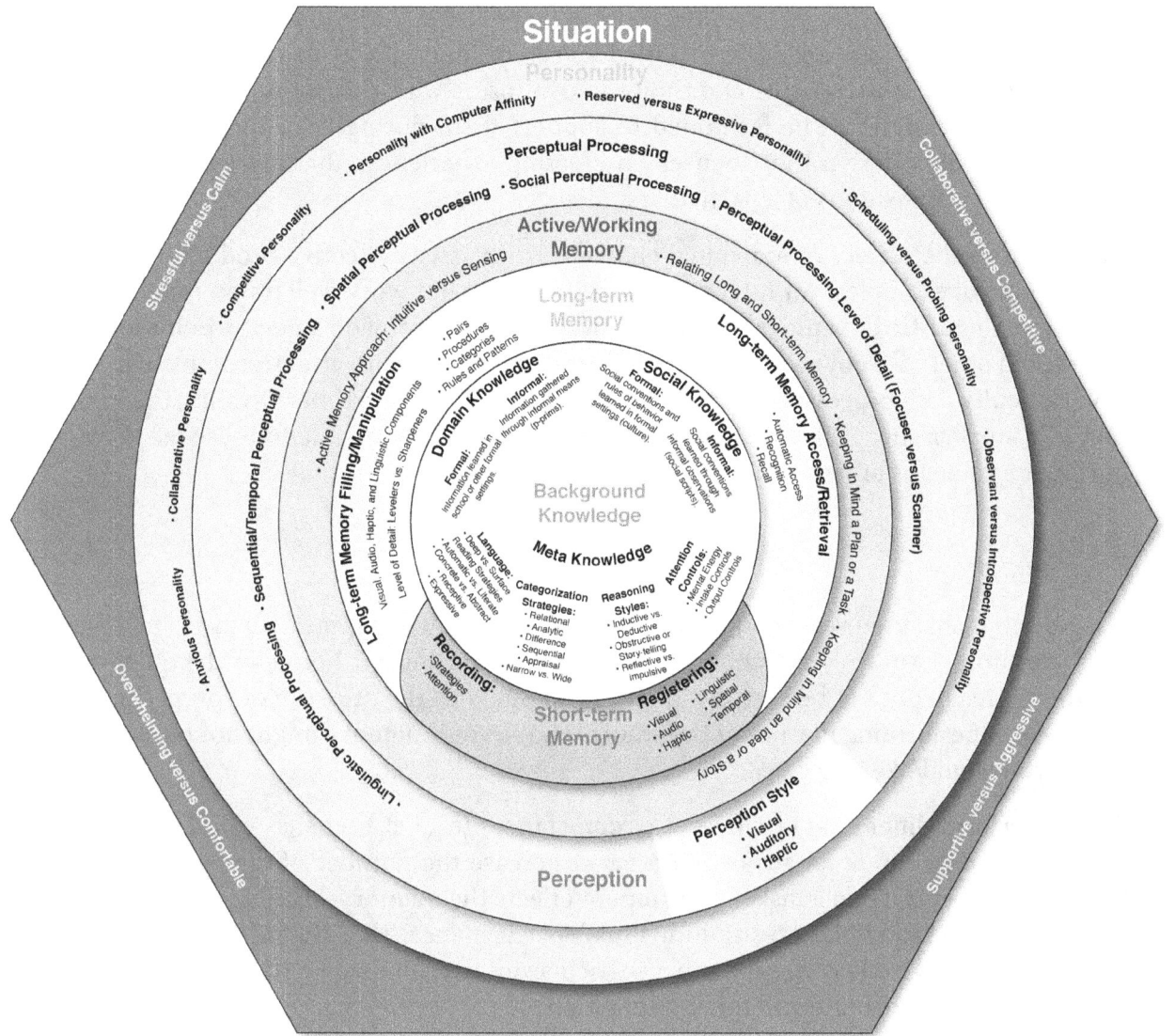

The Cognitive Wheel represents our synthesis of the literature in educational research and personality assessment with an eye towards its use in solving product design problems. Many of the underlying classification taxonomies are discussed in the following chapters with my rationale for the manner in which they have been adapted within the Cognitive Wheel.

Attention controls are discussed separately since they effect all of the above traits. Most of these cognitive attributes are situation-dependent—it's important to consider the setting and circumstances under which the product is to be used. As you can see from this list, it closely matches the variables discussed in relation to optimal experiences—flow. Many of the underlying classification taxonomies are discussed in the following chapters with my rationale for the manner in which they have been adapted within the Cognitive Wheel.

What's most important for a product designer is to be able to identify a particular set of cognitive traits that are relevant to individuals working with a product under a specific conditions. Once the strengths and limitations of users with this set of cognitive traits are understood, a design can be fashioned to support them during their interaction with the product. Product design thus focuses on creating experiences that are well-suited to the needs and predispositions of a particular audience.

Awareness of the variety of individual characteristics assists effective product design. Here are a couple of examples. An Internet chat room might not work well for an individual with expressive linguistic difficulties, but an opinion poll could provide an entry point for people who would otherwise not participate at all. Some individuals with attention control problems have difficulty with processing long pieces of information at one time. Dividing the information into smaller encapsulated chunks makes the material accessible to these users without hurting its value for others.

Environment

Imagine yourself at ten years old, standing on stage in front of a large audience, and trying to spell "accordion" to win yourschool's spelling championship. At home, in the quiet of your bedroom, this might not be such a difficult task. But with the pressure of competition, with the stress of performing in front of the whole school, your memory might not be as cooperative as you would like it to be.

People perform differently in different circumstances (e.g.: at home, at work, alone vs. in a group, in a museum, at an airport, at a bar). Increase the number of distractions and it is harder to concentrate on a task. Environment effects the memory—the anxiety and stress of a particular situation influences an individual's performance. Thus the informational content that an individual is likely to get from a particular interaction is very strongly dependent on the environmental conditions of that interaction.

Social Knowledge

When someone gives you a present, you say "thank you." When an old woman gets on the bus, you give up your seat to her. When you go through airport security, you take off your shoes when asked. When you want to speak in class, you raise your hand. All of these are examples of instances of social knowledge used appropriately in the right situation. There are millions of bits of such knowledge, and all are culturally dependent. When someone screws up, we know it. Computer scientists lament just how difficult it is to teach a robot to perform in socially appropriate ways.

There are two types of social knowledge: formally learned rules of behavior and snippets of social scripts informally acquired. Both types of social knowledge are culturally dependent—they vary between social groups. These differences are not limited to country borders. There is significant variability among the behaviors of individuals belonging to groups defined by language, religion, profession, socioeconomic status, political orientation, and so on. Interaction with the environment is strongly dependent on formal and informal social knowledge.

Memory

There are three kinds of memory storage: **short term memory, working memory**, and **long term memory. Short term memory** allows small bits of information to be readily available for cognitive tasks as they are performed—e.g. keeping in mind a telephone number as it is being dialed or keeping track of items as they are being counted. **Working memory** is where all thinking takes place—e.g. writing requires the author to keep in mind and monitor word spellings, rules of composition and punctuation, handwriting movements, story ideas, and much more. **Long term memory** is what we normally mean when we talk about memory. It is the vast storage of information that we accumulate throughout our lives, and it includes data, procedures, algorithms, and anything else we can think of. To think, we "load" concepts from our long term memory into our working memory and process them together with information from short term memory.

Memory problems can result from insufficient working memory, limited short term memory, and compromised long term memory retrieval—either the information necessary to pass the test is not there at all or it cannot be found. Long term memory stores **passive memories, episodic memories, intentional memories,** and **subject matter knowledge**.

It's amazing how some stuff gets a permanent foothold in our minds without us consciously doing anything. A tune, a phrase, or an image gets memorized without intent, with us acting as a passive vessel. These are **passive memories**.

Intentional memories are the opposite—these are memories that we want to have; information we strive to remember. When we study for a test or learn to use a new device, we explicitly try to commit information to memory. Forming intentional memories requires an act of will and is harder than gathering passive memories.

Episodic memory stores the "movie" of our lives. It's a time-based string of events that creates a record of our daily existence. Events that are colored with strong emotions tend to be easier to bring forward from our long term memory.

Subject matter knowledge is also stored in our long term memory, but its organization is based on our understanding of the subject matter. You can think of subject matter knowledge as ideas, concepts, and rules organized as a structure of information. If an individual has a poor grasp of the subject matter, his structural organization is weak and his ability to retrieve the right bit of information at the right time is compromised. It's the organization of information that creates the most striking contrast between experts and novices. So it's not just how much you know, but how well that information is interconnected.

A particular individual can have a very good episodic memory but have problems with subject matter knowledge. People suffering from poor attention controls tend to also have problems with subject matter knowledge acquisition even when they display brilliant episodic memory.

Domain Knowledge

When we hear "He's a doctor," we presume that man knows a lot about medicine. When we hear "She's a rocket scientist," there's more ambiguity—is she a very smart woman or does she build rockets for a living?

Can you be smart and stupid at the same time? Does the assessment of how smart you are depend on where you come from and with whom you hung out? We certainly have many jokes that relate intelligence and cultural background, most of them in poor taste.

Domain knowledge is a set of expertise and beliefs that an individual has acquired in a particular area (e.g. physics, art history, cooking). An individual's personal assessment of this expertise tends not to be very accurate. And, over the course of a lifetime, a person is likely to acquire a large collection of beliefs that are simply not true and often times contradictory.

Domain knowledge has a lot to do with how we design a product to meet the needs of a particular audience. A professional oven is different from one installed in a private home. It's not just the difference in expertise of the user, it's also the product designer's expectations as to the amount of electrical power available in a professional facility, the output needs of a commercial chef, and the safety equipment built into a bakery but lacking in a family kitchen. And, of course, the goals of a chef are very different from that of a home cook: most us have no desire or need to create hundreds of loaves of bread each day. Thus product design depends not only on what a person knows, but also on the needs of the chosen domain.

Organization of Knowledge

Some kids line up their little car collections in neat little rows, others live in chaos. Some people are never late, others seem always to be just behind the bell. Many individuals suffer in their personal and professional lives from a lack of organization. It's also an easy trait to spot—one glance into a bedroom or an office, one missed appointment, and...

Those born with a gift of organization and scheduling skills have a much easier time sailing through the rules and regulations so strictly imposed by the large bureaucracies of our world. These individuals go on to become information architects and librarians! But for those others, for whom time, resources, and space management are challenging, "getting it together" is a painful proposition and not easily accomplished. These individuals need help, and most can't do it on their own.

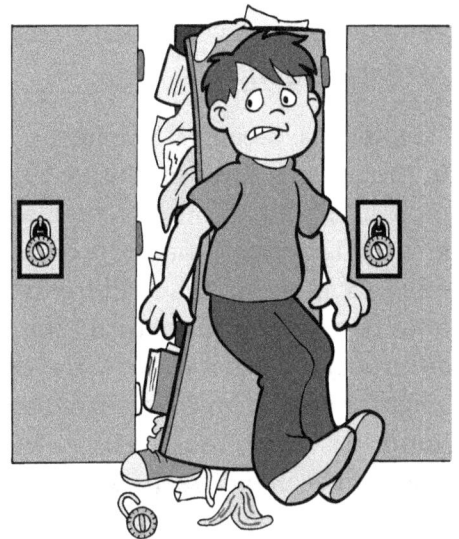

Disorganization goes deeper than mismatched pairs of socks or the buried surface of a desk at work. Disorganization applies to thinking as well. The same skills necessary to organize a closet are needed to organize an essay or a financial report. Some individuals manage to keep their computer files in neat order, while being unable to keep physical objects that way. Some are just the other way around. Both types need support structures to help them maintain order in those tasks that they find difficult.

As product designers, we can make it our goal to teach these people time management and organization skills. But this is not really part of our job description, is it? Yet if we develop products that our users fail to use, then the incentive to do something about it is very high. We have to build-in help for these individuals. We need to provide cognitive scaffoldings.

Perception

Human perception is an information gathering tool. Perception can be separated into acquisition and processing. There is the information that an individual receives from his senses—the visual, audio, tactile, and other perceptual information cues derived from the environment—that's acquisition. This information is then processed by the brain. An individual can have a perfect eyesight and still have problems decoding written language—perfect acquisition of visual information, but faulty processing of linguistic content. There are also individual variations in perceptual preference: some people rely more on their sense of vision and others on their hearing.

Perception is strongly influenced by environmental conditions (e.g. low light, crowding, extreme noise, stress), by domain background knowledge (e.g. knowing what to pay attention to in a given situation, knowing what's important), by social knowledge (e.g. religious restrictions on observation, manners), and by attention controls (e.g. low mental energy, impulsivity, hyperactivity, hyper focus, insatiability).

Attention Controls

"She is a daydreamer. She never pays attention to the teacher. She's impulsive." These are common utterances from teachers and parents of kids with attention control problems. But there's more to attention than failure to sit still for an hour or speaking out of turn. Attention controls govern our behavior. They are responsible for what we notice, how we relate new experiences to our past, how much energy we have to do a task, and even our sleep patterns.

There are three categories of attention controls: **mental energy** controls, **input** controls, and **output** controls. **Mental energy controls** describe the differences between various people's ability to focus intensively on a task for a period of time. Some individuals need a lot of little breaks while they perform cognitively difficult tasks, some don't. Some people sleep well at night and some don't. Everyone is different.

Intake controls describe individual differences acquiring new information. For example, some students are seemingly born knowing how to study, others have to learn how. Some individuals understand what's important and can focus on just the right ideas for just the right amount of time. Others have more difficulties finding the informational cues. Some people get enormous gratification out of getting every problem right, others don't find much pleasure from the effort.

Output controls describe the differences in personal work product. The same mental effort can result in a prolific output or a scant dribbling of work. Some individuals talk about working but produce very little. Just spending a lot of time is not enough. Clearly, attention controls have an enormous effect on how a particular person or a group interacts with a product or functions in a particular setting.

Language

It seems obvious that language would effect product design—instructions alone can make all the difference in the world. But language is far more integrated into design: language skills are critical to successful product use. Inability to express one's ideas clearly doesn't only limit an individual's ability to communicate with the world (in both written and oral form), it hampers comprehension. Problem solving, in any subject matter, requires internal dialogue—we "talk" ourselves through a problem. And poor language skills result in poor listening skills and lead to impoverished understanding. Poor language skills don't only manifest themselves in Language-Arts classes, they hamper learning and work across all subject areas.

Additionally, poor language skills can be the cause of impulsive behavior and poor emotional control. If you can't think about a problem, you do the first thing that comes to mind. And reasoning—talking to yourself about your feelings—helps stabilize emotions.

Product designers have to understand the language limitations of their users. And in world where product use rarely takes in account country borders, it's not only one language you have to worry about. And it's not only the users' language skills, it's also the designers' comprehension and communication that we need to consider.

Personality

Some people can take a bad situation and really drive it into the ground. Others drip with charisma—they seem to be able to get out of any trouble without a ding on their self-esteem or a scratch on their dignity. Personalities can either burden us or serve as a free ticket. Most of us fall somewhere in-between. As adults, we can choose professions that best fit our personalities. As product users, we are stuck within a design system that is not very forgiving of social variability. On top of this, users have to cope with other users, and cruelty runs rampant.

Social problems come in several flavours. Some users of online social group, for example, are ostracized and have difficulties acquiring and keeping friends. Others have difficulties disentangling themselves from their social lives and have problems producing any work due to social demands. Both conditions result in poor performance, but require quite different product design solutions.

High anxiety is another personality trait that can lead to loss of productivity and even result in serious physical symptoms (e.g. stomach aches, headaches, vomiting, hyperventilating, sleep disorders, etc.). Incorporating anxiety management techniques into the design of a product (e.g. manuals and instructions, clear information architecture, etc.) can greatly reduce the symptoms and lead to improved product performance.

Some individuals exhibit disruptive or even destructive behavior, damaging themselves and those around them. These behaviors are usually symptoms of other problems. As those other issues are addressed, the undesirable behavior tends to subside. Product design can limit the damage from such individuals by building-in constraints on usability.

Friends that people choose have a lot of influence over how they behave and the decisions they make. Individuals develop reputations among their peers. The reputation sets expectations on behavior and performance, both at work and with friends. Once the reputation is developed, it's very difficult to shake it off—it's very difficult to change how others perceive you and what they demand of you. If the boss sees you as a slacker, she will treat you as a slacker even when you might be trying to improve your performance. Changing that opinion is orders of magnitude more difficult than getting it in a first place.

The same is true for the social circle of friends—once a group classifies a member (rightly or wrongly), it will push that person to conform to its expectations, including pushing this person into disruptive and destructive pattern of behaviors, far surpassing those that the person is capable on his or her own.

Anorexics form groups that glorify starvation as a life style. They describe their behavior on the Web as "praying to the Goddess Ana." Someone who is just leaning in the direction of dangerous body image misperception, can be pushed over the edge by associating and socializing with such groups.

Similarly, MySpace can be a supportive environment or a place for cyber bullying, reportedly so severe that it has even driving some members to suicide. Reputation is a very powerful tool. While some individuals are able to overcome a bad reputation, others drown in it. Reputation can poison the environment. How much is MySpace liable for the death of its member? Can product designers limit such behaviors by restructuring the parameters for this site's design?

A Quick Note About Genetic Intelligence

Genetic intelligence is composed of reasoning abilities that an individual is born with. By definition, it is not situational and does not change over a person's lifetime.

Genetic intelligence describes adaptive mental behavior in unfamiliar situations. Genetic intelligence represents different forms of reasoning including abstracting, forming and using concepts (classification); perceiving and using relations; identifying correlates; maintaining awareness in reasoning; and abstracting ideas, especially from figural and nonverbal, symbolic and semantic content. Genetic intelligence

is non-mutable, independent of content, and culture free. Genetic intelligence is more influenced by biological factors, such as heredity and central nervous system growth and maintenance, which is consistent with other culture-free aspects of heredity.

Genetic intelligence is not represented in the Cognitive Wheel. If you encounter a design situation where you are faced with users of below or above average intelligence, you will still need to understand the limitations and/or strengths of that audience and address them accordingly.

The Brain

All of us get comfortable with the locations of our fingers, knees, elbows and such pretty early in life. Very few of us get to know our most cherished organ, the brain. While the following illustration isn't good enough to perform surgery, it does shed a bit of light on the structure of the brain and where the parts which perform some of the different functions we are discussing are located.

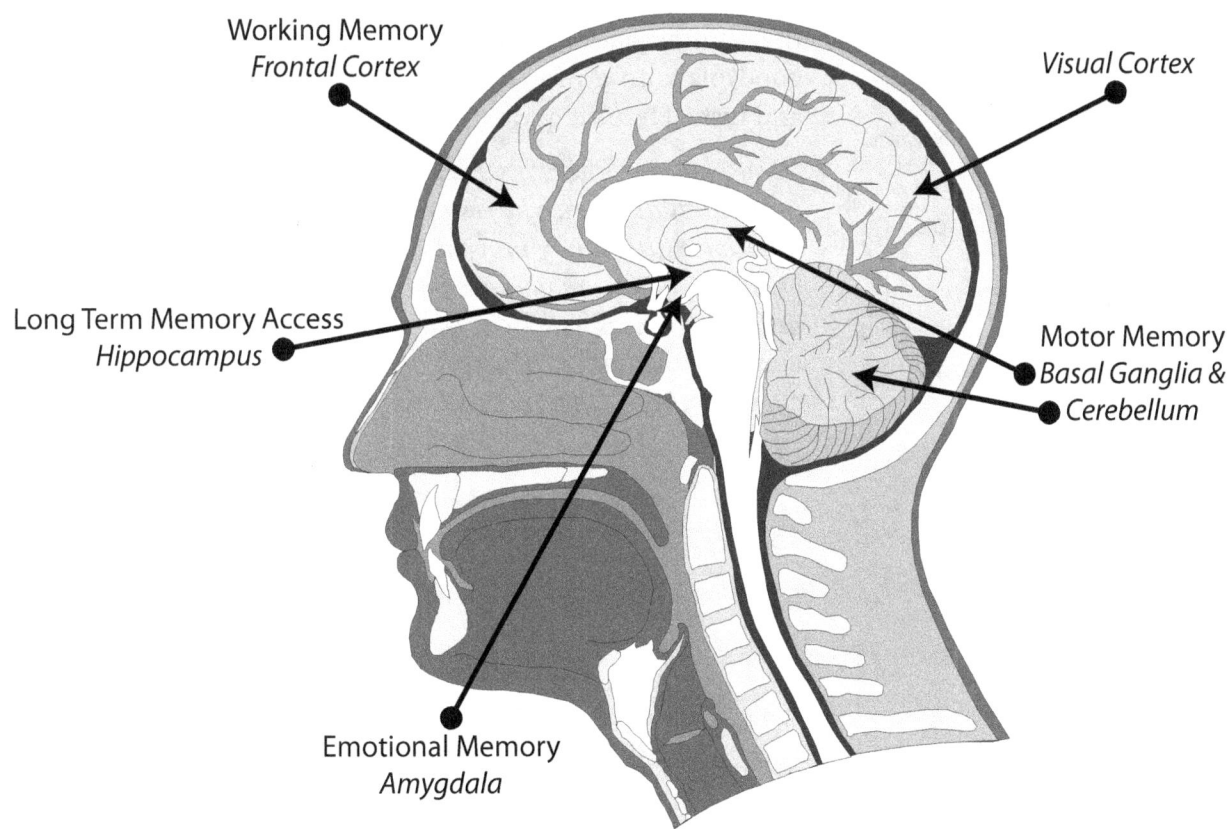

Additional Thoughts and Further Readings

In the movie, "The Gods Must Be Crazy," (1980) a native of the Kalahari Desert finds an empty glass Coke bottle—an unknown technology to his people. The story documents the misadventures of people using a novel device and of a culture that doesn't throw away such valuable objects. It's an entertaining illustration of a product used in ways that were not predicted by its designers.

This chapter divides cognition into background knowledge, memory, perception, attention controls, and personality. Each of these traits will be discussed in subsequent chapters. Environment has a strong influence on how we perceive the world around us, what we remember, and how flexible we are in using the products we need.

The language section of this book provides more details about parts of language and their effect on language skills.

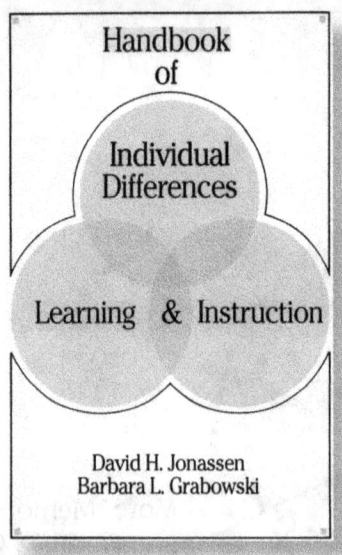

David Jonassen and Barbara Grabowski wrote a textbook on psychometrics (how to identify and test for cognitive differences): "Handbook of Individual Differences, Learning & Instruction." It summarizes the research in psychology, cognitive science, and education. It also provides educational interventions for various cognitive deficits. This is a good place to go for quick overviews of different theories and tests. Here's a quote:

"Every learner filters instruction through a set of individual difference filters or lenses. Individual difference filters may prevent the mental assimilation or accommodation of ideas by the learner. Individual difference lenses will focus the skills and content in ways that will affect how any individual learns. The effects of those differences are universal."

6. Memory Metaphors

Everyone complains of his memory, and no one complains of his judgment.

—La Rochefoucauld

You can think of memory as a storage space for information; this is a common metaphor for memory in our culture. There are three types of memory "storage": **short term memory**, **working memory**, and **long term memory**. **Short term memory** is a temporary storage for data needed during the task at hand but not meant to be remembered a few minutes later—e.g. keeping in mind the total purchase price as you figure out change. **Working memory** is thinking space—it's where you do the math to figure out the necessary change and also where you stress over making the purchase. **Long term memory** is where we store information that we're not currently thinking about. When people complain of poor memory, they usually mean they have a hard time getting data out of that long term storage.

There is some variation in the abilities of individuals to access and store information in their memories. There are also differences that are due to age. And working memory in particular is very situational. For example, anxiety can reduce a person's memory management ability. Surprisingly, pain can enhance it.

It is important to note that different people have varying capacities for remembering disparate types of information. Some have better visual memory and some perform well with audio information. Furthermore, it is easier to remember things that are of interest to you than information from a subject that bores you.

Memory Components:

- **Long Term Memory**
- **Short Term Memory**
- **Working Memory**
- **Episodic Memory**
- **Subject Matter Knowledge**
- **Intentional Memories**
- **Passive Memories**

Long Term Memory

An education isn't how much you have committed to memory, or even how much you know. It's being able to differentiate between what you know and what you don't.

— Anatole France

Long term memory contains the "movie" of your life; all the social customs and traditions passed on to you by your family, friends, and television; and all the stuff you learned along the way. All that stuff is in there somewhere in your head; the problems arise when you try to get at that information. If we had perfect access to our long term memories, we would rarely have to do any math—we would just remember the solution to the problem at hand from the last time we saw it (one day, one month, one year, or 25 years ago). For most of us, our long term memories are just not that accessible.

A good way to think about long term memory is by making an analogy to a library—it's all in there somewhere, we just need a good index card system and a helpful librarian to guide us to the right bit of information at the right time. So the trick to long term memory is a good retrieval system and, consequently, a great filing system.

Unfortunately, people are not born with a great filing system. It's developed through experience and learning. We need to learn how to categorize the various items in our long term memory. The more we know about a particular subject area, the more flexibility we have creating categories and noticing patterns. Long term memory retrieval is highly dependent on the information's organization. And the more expertise we have in a particular area, the better our categories.

For example, try to name as many movies as you can. As you start listing off the titles, notice if you did it in some sort of a pattern. Chances are, you have a categorization scheme: all the movies by a particular director; or all the films with a certain movie star; or it's a pattern based of genres. Whatever it is, you probably managed to name more than just a half a dozen movies. But ask a young child to do the same task and the answer will look different. Even though most children five years of age have seen many movies (think of the Disney repertoire), they haven't grouped them into good categories yet and thus have trouble retrieving the names from long term memory.

Try the same problem but instead of movies, name different animals. This seems like it will be an easier problem for a five year old, but again, most kids of kindergarten age will not come up with many animals. Their answers will look something like: cow, tiger, chicken, cat. But ask a fourth grader to do the same problem and you will get something like: cow, chicken, sheep, horse; dolphin, whale, seal; shark, tuna, crab, lobster; tiger, lion, giraffe, bear. Both answers have about four chunks, but while for a younger child those chunks represent individual animals, for an older one they are categories of animals (barn animals, ocean mammals, underwater animals, zoo animals). If such memory a categorization scheme isn't evident by the end of elementary school, a child might have long term memory problems that need to be addressed by a professional.

Long term memory retrieval is highly dependent on the organization of information stored within. And the more expertise we have in a particular area of study or a field of interest, the better our categories. For more information about categorization strategies, see the discussion Chapter 9: "Background Knowledge."

"File these in random order like you usually do."

Short Term Memory

So what is short term memory? In 1956, George Miller wrote a paper about his experimental work on the ability of humans to remember information. Based on his data, Miller concluded that people could absorb about seven chunks of information, give or take two chunks, and hold it in their short term memory for about ten to thirty seconds. After that time, if the information hasn't been committed to long term memory, people can't recall that information correctly. (Note: Dr. Mel Levine believes the time to be more like 5 to 10 seconds or less.)

What constitutes one chunk of information depends on how well a person knows the subject area. Most English speaking people can remember seven numbers (as long as those numbers are between 0 and 100), seven English letters, seven English words, and even seven English phrases if they are already familiar to them (like a movie or a song title). But if an English speaking individual is asked to remember Chinese characters, for example, that number decreases significantly. The less familiar the subject matter or task is to a person, the less likely it will be remembered. Short term memory capacity also seems to decrease with age, so that an older individual can hold a smaller number of chunks in short term memory than a younger person.

It has been shown that novices chunk information in smaller pieces than experts performing the same tasks. For example, an expert Photoshop artist working in that computer application would need far fewer directions to accomplish a specified unfamiliar graphic arts task than a Photoshop novice—the size of each "chunk" is larger for an expert than for a novice.

Options at the top of the search page show options for different types of searchers: Web, Images, Maps, News, and so on.

Options just below the search box suggest the narrowing of search by focusing on the type of information the user is looking for: Web, Books, Video, Scholar, and Groups.

The search term (or search expression) is always available. "Advance Search Preferences" help step the user through a process of narrowing the search.

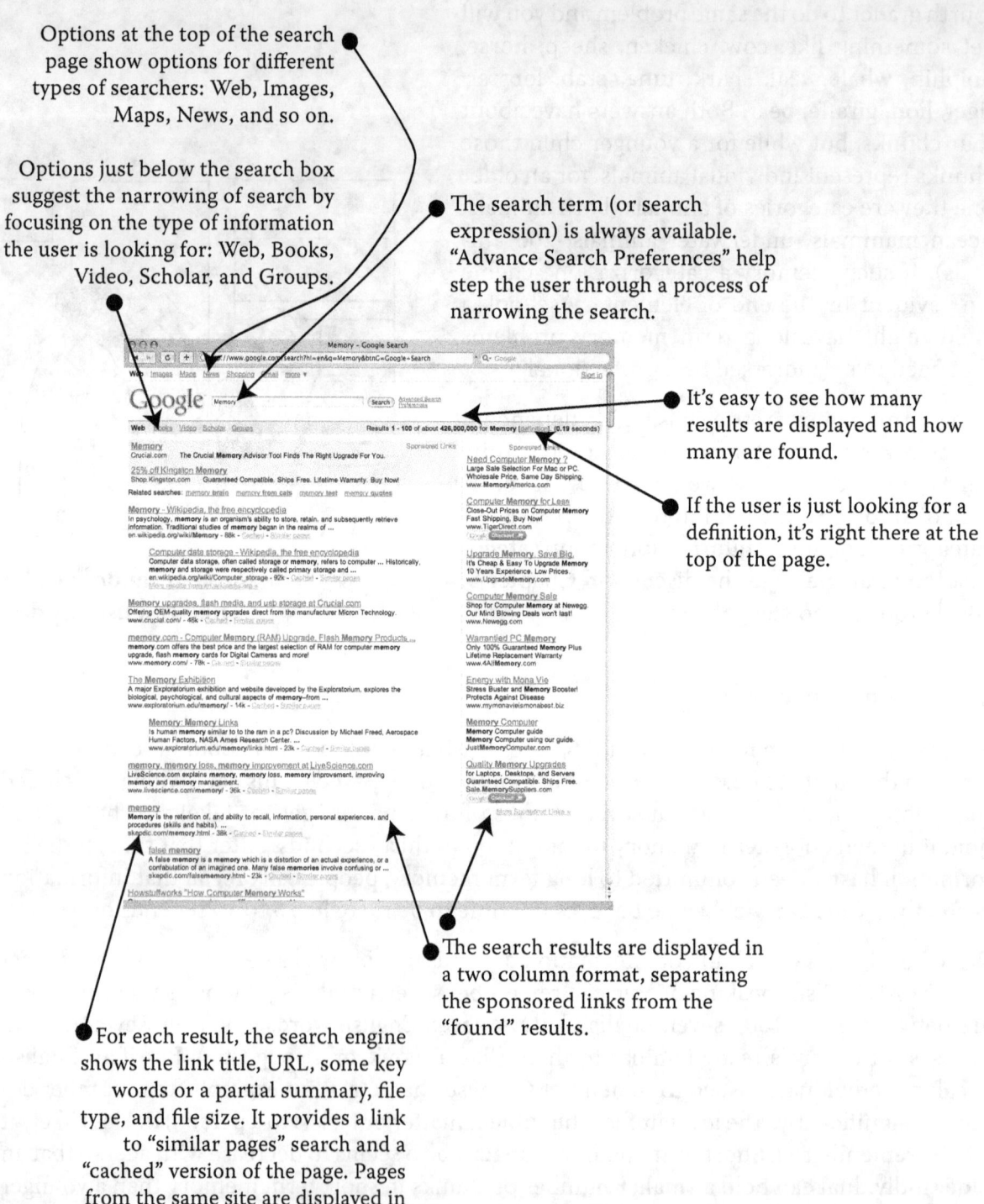

It's easy to see how many results are displayed and how many are found.

If the user is just looking for a definition, it's right there at the top of the page.

The search results are displayed in a two column format, separating the sponsored links from the "found" results.

For each result, the search engine shows the link title, URL, some key words or a partial summary, file type, and file size. It provides a link to "similar pages" search and a "cached" version of the page. Pages from the same site are displayed in tabbed arrangement.

Another example is the ability to remember a piece of music. An expert might be able to replay an unfamiliar piece of music by just hearing it once, while a novice might have trouble even remembering the first few notes. While this seems an obvious distinction, it's critical to understand when designing products for a specific audience.

Perhaps an example of how this memory limitation is often ignored will make clear why it needs to be addressed. Search engines are currently set up on a keyword-based search system. After a boolean search has been executed, the user is presented with a large list of "hits" which are prioritized using that search engine's proprietary algorithm. The user then must sift through that list to find the results that are actually relevant to the user's query. But this activity is constantly interrupted—there are many times that the user has to visit a site to determine if it is relevant to her search. Visiting and determining the value of a site is a significantly different activity from going through a list of results. And after a user has jumped to a specific site, she may begin sifting through the content there trying to find information relevant to the search. What usually happens is that the user becomes distracted and finds fascinating information that isn't relevant to her original query. And every time she has to step back from the search results and jump into a suggested site, she loses track of the results she wanted to check out next.

There are many "tricks" the search engines use to help their users improve their productivity during search tasks, but some people still manage to waste enormous amounts of time taking side roads and exploring the unexpected gems revealed during this exercise. Search trekking is a hobby of many students conducting research for their school papers, and it is also a time sink for those happy to let their attention wander through cyber space.

One obvious solution to improving working memory management during a search is to implement a tagging system for the search results. The results that seem promising can be tagged and rated by the searcher and saved for later review. This simple approach would keep the tasks of sifting through the search engine results and analyzing them distinct, resulting in uninterrupted flow of activity.

Short Term Memory Test

Short term memory comes in different "flavors"—visual, haptic (taste, touch, muscle motor memories, and proprioception—orientation of the body in space), audio, and linguistic/ symbolic.

To test your symbolic short term memory, try repeating a sequence of random numbers given to you by a friend after a 10 second delay. Continue to increase the number of numbers (use a new set of numbers at each try, otherwise the sequence gets committed to long term memory). Do not use telephone numbers—they are socially defined patterns that we are experts at remembering. You can do the short term memory test with music, sounds, or graphical information. The less you know about the subject matter, the more limited your short term memory is in that context. Try adding distractions like noise or carrying on a conversation during the test. What is the impact of the environment on your short term memory?

To test your visual short term memory, turn to the Appendix and look at the Short Term Memory Test illustration for no more than four seconds. Wait a dozen seconds and then draw that illustration from memory in the box. Give yourself no more than a minute to create the drawing. When done, compare your drawing with the original illustration.

Did you get all the elements? Are they in the right position? What are the differences between your drawing and the illustration?

Working Memory Definition and Working Desk Metaphor

Working memory is allocated to understanding and processing information, and to problem solving. A good way to think about working memory is to visualize it as a work desk. Using the same metaphor, think of a short term memory as objects that can be placed on that desk. Long term memory, continuing our metaphor from before, becomes a library of information with books that can be examined on this work desk.

For example, while doing an addition problem, all the numbers that need to be added together, along with all the interim solutions, are stored in short term memory, while actual addition happens in working memory.

We all have a unique size working desk—some desks are smaller, and some are larger than average. The size of each person's desk is relatively stable—it is hard to stretch the desk to change its capacity. Once the desk is filled (one layer thin), no more additional information can be put on top without other bits falling off. This is important—there are strict limits to how much stuff we can hold in mind and manipulate at any one time, and that limit is very limited indeed.

If an individual is tired or presented with an unfamiliar task that makes him nervous, or if an individual suffers from anxiety, the performance capacities of both short term and working memories decrease. That memory—that space on the working desk—is being spent on worrying or focusing on other things rather than the task itself. Emotion takes up working memory space.

Working Memory Test

To test your working memory, try playing "Simon Says" or the "Memory Game" (you can use cards and match reds with reds and blacks with blacks). A standard test is to give the subject a random list of two-digit numbers and then have the subject repeat the list in reverse order. Since they have to manipulate the numbers prior to reporting them, it is a fair test of their working memory capacity. Again, try adding environmental distractions and see the effects.

One experiment I performed in a fourth grade class was to have a student perform multiplication problems: first in a quiet classroom, and then with two other kids shouting numbers in his ears. As you might expect, the performance fell drastically during the second half of the experiment. But he was still capable of getting the right answers standing in front of the whole class in the first half of the experiment. Another student could not—her working memory was taken up with performance anxiety. It's not that she wasn't able to do the problems at all, she just couldn't in front of all those people. We are all familiar with this memory trap—we get so worked up about what we need to do that there are no resources left to actually do it. That's why certain people do so poorly on tests—they have never learned how to clear their working memory of extraneous thoughts.

Working Memory and Personality

Working memory is related to our personalities, although the causality is not all that clear. It could be that an individual's propensity for observing detail limits his ability to keep other information in mind. But it seems equally possible that some people's working memory is better at storing observational data. For product design purposes, the causality is not very important, but the differences matter a lot.

The personality traits, **introspective** and **observant**, relate to how we allocate our working memory resources: observant people have more working memory allocated to observation (sensory information) and introspective people have more working memory allocated to internal dialogue (internal flow of information). A good way to think about this difference is by using the working desk metaphor again: how is the space on the desk divided between information that is coming in from the senses and the stream of thoughts generated introspectively? There is a very limited space on the table and how we divide that resource is a personality preference.

In a similar matter, the **emotional** to **reasoning** personality continuum impacts memory allocation. Everybody has both thoughts and feelings but working memory limitations mean that it is difficult to pay attention to both at the same time. By definition, therefore, an individual cannot be high on both the Emotional and Reasoning scale at the same time. A person can pay attention more to one or to the other at any given instant.

There's more about personality differences and their impact on product design later in this book.

Additional Thoughts and Further Readings

> *It has been remarked that the very essence of civilization consists of purposely building monuments so as not to forget.*
>
> —L. S. Vygotsky

George Miller's seminal and much-quoted paper, "The Magical Number Seven, Plus or Minus Two: Some Limits on Our Capacity for Processing Information," was published in the 1956 issue of the magazine Psychological Science. "The Magical Number Seven" has entered our collective cultural vocabulary—a feat few experimental outcomes have managed to accomplish.

Mel Levine's book, "A Mind at a Time," is inspiring—we seem to know quite a lot about things that go wrong neurologically in the brain, but so little of that diagnostic and medical knowledge actually gets applied in the classroom, or in the field of product design, for that matter.

Dr. Levine is a neurologist, a medical doctor who treats children with neurodevelopmental problems. He is a

A MIND AT A TIME

AMERICA'S TOP LEARNING EXPERT SHOWS HOW EVERY CHILD CAN SUCCEED

MEL LEVINE, M.D.
FOUNDER, ALL KINDS OF MINDS INSTITUTE, and DIRECTOR, CENTER FOR DEVELOPMENT AND LEARNING

"narrow categorizer" (he admits as much)—his tendency is to relate to each child in his medical care as a unique case. This is very good approach if you are a doctor. My interest is in product development, including curriculum design—by definition, this makes me a "wide categorizer." But his book inspires readers to come up with instructional strategies that would help individual students with particular neurological handicaps. And in the process, Levine argues, these same strategies will work for the majority of the students in a classroom. So while he is advocating individualized curriculum at some level, Levine is indicating that the adoption of the same instructional strategies that help students with learning difficulties to succeed will also work to improve the overall level of instruction in the general classroom. The same strategies are also be of use to product designers.

If you find this subject as fascinating as I do, I recommend reading an actual textbook on educational psychology. There are many to chose from. I liked Anita Woolfolk's book and read her seventh edition of "Educational Psychology." The current edition number is up to number ten.

For more information about working memory allocation due to personality differences, please read Chapter 11: "Personality" later in of this book.

And here are a few questions. Do you think there is a cultural component to working memory allocation? Do some cultures prefer individuals that are more observant? Do we encourage introspective personalities? How do different cultures feel about emotional as opposed to reasoning people? Do these attitudes change over time?

8. Long Term Memory Filing and Manipulation

When you know a thing, hold that you know it; when you know not a thing, allow that you know it not; this is knowledge.

—Confucius

Long term memory stores **passive memories**, **episodic memories**, **intentional memories,** and **subject matter knowledge**. These memories can contain visual information, audio data, linguistic thoughts, symbolic ideas, and haptic knowledge.

A library metaphor is a good way to conceptualize long term memory. All the "bits of data" that we have gathered over our lifetimes are stored somewhere in our cortex-like books on library shelves. Our hippocampus acts as a card catalog and a librarian all rolled into one little bit of gray matter. No hippocampus, no access to long term memory—individuals who have lost theirs through disease are not only unable to retrieve their long term memories, but also can't form new ones.

Many were introduced to short term memory in Pixar's 2003 animated film "Finding Nemo." In the story, the character Dory is said to have a short term memory problem. In fact, as presented in the movie, Dory's problem is storing new memories into her long term memory. It's not that she can't manipulate ideas in her head, it's that she can't remember what she had done just a few minutes prior to the present. This echoes the more serious film treatment of the same malady in the "Memento" (2000).

Visual, Audio, Haptic, and Linguistic/Symbolic Memory

Look at the illustration above. The product containers are rendered as rough silhouettes, yet any American can easily guess what these products are. This is our visual long term memory at work. We can do similar feats of long term memory remembering snippets of music, the characteristic bounce of our partner's walk, the smell of grandma's cookies, the feel of sand between our toes, and the words of our favorite poem.

Not everyone is equally good at remembering the different "flavors" of memory. Some are particularly good at remembering audio-based information, for others visual memory works best. Dancers and athletes are very good at motor memory.

Actors develop abilities for incredible linguistic memory feats—they are able to store pages of script in a minimum amount of time (sometimes with just two or three read-throughs of dialogue). Clearly, these individuals weren't born to remember lines, they developed this skill over time in the course of their professional lives. But it helps to have a natural predisposition to remembering this kind of linguistic information.

Yet our memory is very faulty. It is heavily colored by our beliefs, culture, and experience. Most police officers and trial lawyers know just how unreliable eyewitness accounts can be.

The illustration of silhouettes of containers above

seems easy for an American, but might be completely incomprehensible to someone from a different cultural background. Visual symbols are cultural items designed to trigger recognition of an idea or concept stored in long term memory. If a particular symbol doesn't trigger recognition, it is either poorly abstracted or too culturally specific to be recognized by members of different cultures. This is something to keep in mind when designing iconic images for multinational audiences.

Emotional Memory

Just like we can remember color, shape, idea, or a rule, we can also recall emotional information: "It was a warm, sunny, summer day and I felt happy." Not only do we remember feelings, but heightened emotions serve as triggers for other memories. It's easier to recall an event that is embedded into strong emotional context: the death of a friend, a car accident, a stay at the hospital with a sick loved one, a wedding, a first kiss, the birth of a child. All of these events are charged with emotion and the details are chiseled into our long term memories. And negative feelings are just as good at creating these easily retrievable memories as happy ones.

This association between feelings and memory is well known to book authors and film makers, advertising gurus and museum exhibit designers. In the museum design trade, exhibitors know that they can cement memories by evoking an emotional response. You walk along a narrow, dark, curving corridor and suddenly come upon a giant, bright, open room with a full scale T-rex. That's such a moment. The feelings of surprise and excitement help create vivid memories of the event, and these memories will jump to the front when the visitor thinks back about that museum outing.

Automatic Access

Automatic access from long term memory occurs when two or more bits of information are strongly linked in memory such that when one bit comes up the others immediately come to mind. Consider a multiplication table: five time five is twenty five; six times seven is forty two; etc. There was a time early in our elementary school careers when multiplying two single digit numbers was very hard—we actually had to do some multiplication! But after a lot of drill and practice, the numbers and their multiplication product are very strongly linked in our memories, so that the answer just pops up when we need it. This is an example of automatic access.

The advantage of automatic access is that very little working memory (space on the desk) is spent on deriving an answer. A high school student presented with a complicated quadratic equation will need most of that working memory solving an unfamiliar problem rather than wasting it on multiplying simple numbers together. Now consider a pilot—this job requires a lot of fast thinking in times of emergency. Imagine if a pilot had to actively look for an altitude display or stop to consider whether pulling back on the stick means going up or down. In an emergency, the position of instruments and the use of the controls has to be like second nature to a pilot—there shouldn't be any waste of working memory on trivial details. All of a pilot's processing power needs to be directed to solving the emergency.

While automatic access is the most effective at clearing the space of the working desk, it is also the hardest to achieve. Again, think back to learning multiplication tables—it was time consuming, and drill and practice sessions are rarely fun. But automatic access to information is one of the characteristics that define an expert in a field. Most of us are multiplication table masters, having achieved automatic access to the results of simple multiplication problems.

Automaticity takes time. Just like a perfect golf swing takes many hours of practice, automaticity of language, math, and product-using skills is not achieved overnight.

Recognition

"I don't remember, but I'll know it when I see it." This is what recognition is all about. Many multiple choice exams are tests of recognition. GUI interfaces are based on the recognition skills of their users—that's why they're easier than command line language interfaces. As product designers, we strive for recognition.

Cold Recall

Cold recall is the hardest of all long term memory access and retrieval procedures. It involves pulling out information from the long term memory without any external support structures at all.

When US citizens are asked to give their social security numbers, they are asked to recall a nine digit number without any assistance. Fortunately, most Americans have invested the time and mental energy to be able to do this extraordinary feat of memory, achieving automaticity.

Cold recall is the hardest memory retrieval task and one demanded on a regular basis at school. The classic case is studying for a history test that will require knowing many dates, people, and events. People with poor memories suffer most in situations where they're subjected to cold recall.

Memory Filing: Pairs, Procedures, Categories, Rules, and Patterns

Most people have a "to be filed" folder—a collection of miscellaneous papers, letters, and documents that are too precious to throw out but too hard to properly file. To find an item in this folder takes a long time—we have to look through each paper. The initial investment into a good filing and index systems always seems like a good idea when some desired paper is just not turning up in the "to be filed" folder.

Our long term memory has similar issues: how can we pull out just the right bit of information at the right time? As I discussed earlier, our working memory is of limited capacity—pull too much information from long term memory, fill up the work desk, and items just start to slide off that desk. In a well organized memory, the information is stored and indexed to make accessing it easier—just a few items to look through as opposed to browsing through the entire "to be filed" folder. Such organization of long term memory comes with expertise—the more you know about some subject, the better that knowledge is organized and the easier it is to pull up.

Information in long term memory is stored in several different ways: in pairs, procedures, categories, rules, and patterns. **Pairs** are linked pieces of information: a name with the face of an individual; a word with its spelling; an entry in a multiplication table—two numbers and an answer. When a data pair has a strong link, the retrieval becomes **automatic**—precious working memory space doesn't have to be used on trivial processing. Students that don't achieve automaticity on simple tasks like spelling or multiplication, have their capacity to do more difficult work like writing essays and solving quadratic equations severely diminished.

Procedures are like recipes: how to do long division; how to look up a word in a dictionary; how to write a research paper. When procedures are not automatic, students have to think about them—how does long division work?—instead of actually using them. Since

Long Term Memory:

- **Pairs**
- **Procedures**
- **Categories**
- **Rules**
- **Patterns**

working memory is limited, the more that's taken up with thoughts about procedures, the less there is to think about a problem.

Patterns are visual, audio, haptic, linguistic, symbolic, temporal, spatial, or sequential similarities among data. A child with a good eye for patterns doesn't only excel in the game of "Set," she also has an advantage when it comes to creating categories for grouping information.

Categories are groups of similar bits of information. Without making categories and taxonomies out of the items in the world around us, it would be very difficult to remember or learn anything. What are the types of items that you can buy in supermarket versus a hardware store? What are the nutritional characteristics of the foods we eat? What remedies can treat various maladies? We organize our world continuously. We look for patterns and classify objects and people around us. Forming categories and sorting information into them is our way of organizing our long term memory—creating our good librarian that can access the stores of data we have stashed away in our heads. The better the categories, the easier it is to retrieve the right information at the right time. But if data is miscategorized, a student might erroneously apply the wrong formula to a math problem, for example.

This ability to create categories and fill them with different ideas and concepts develops over time. Consider our earlier example about how a four year old might respond to being asked how many animals he can name. The kid might respond something like: "Cow, dog, monkey, fish, sheep, duck." A ten year old child asked the same question, will exhibit better organization of her long term memory by coming up with: "Cow, sheep, chicken; tiger, elephant, gorilla, giraffe; whale, dolphin, seal." Both the younger and older child can name quite a few animals. But in a case of the older child, the answers will show a categorization strategy at work. Kids have to learn how to form effective categories and they need to develop strategies to form them.

Rules are sets of instructions which surround a topic. Think, for example, of the rules of English grammar. Our society creates institutions that revel in rules: rules of behavior, rules for doing work, rules that govern driving, rules of logic, and so on. Some rules are arbitrary (e.g. we must always drive the car on the right side of the street in America, but not in England) and thus impossible to derive through logic. These just have to be memorised by every individual.

Rules and Culture

The public is wonderfully tolerant. It forgives everything except genius.

—Oscar Wilde

When it comes to cultural differences, social rules of behavior are the most problematic. While morality and the basic code of conduct based on it are fairly uniform—thou shall not kill; thou shall not steal—the laws adapted to enforce the moral code vary from country to country and culture to culture.

We know the rules of our society because we grew up soaking them in—we've undergone gradual indoctrination to the social norms of the culture in which we live. For the outsiders, some of our rules might seem strange or even silly. But for the insiders, they are a natural set of conditions and limitations developed for the good of whole.

Rules set expectations on behavior—because I understand the rules, I will understand the motivations of others in my social group and will be able to predict what they will do in a particular situation. If my expectations are not met, I will feel anxiety and my ability to perform well on a task will diminish. Culture shock is in part a result of not knowing the rules and inability to form expectations about motivations of others.

For product designers working on developing multicultural projects, the ability to correctly set the expectations of their users is a key to product's success. This requires intimate knowledge cultural rules. While television and movies have helped to create a more uniform set of expectation between cultures, the cultural rules they tend to spread are American. Ethnographic information about potential users has to be obtained and applied during the conceptual, interaction, and interface designs phases of product development.

Intentional vs. Passive Memories and Episodic vs. Subject Matter Knowledge

Passive memories don't require effort, they just manage to lodge in our brains. **Intentional** memories we have to work at, to memorize on purpose. Passive memories do not require a decision, while intentional memories are formed through an act of will: "I will commit this information to memory."

Episodic memories are a time-based series of events stored in long term memory. **Subject matter knowledge** is information organized by topics, and it also resides in long term memory. You can think of subject matter knowledge as a relational database—relationships between ideas, concepts, and rules define the structure of the organization of knowledge. If you consider the culinary subject matter knowledge of a chef versus an occasional cook, the difference wouldn't be limited to the amount of overall knowledge. It's the organization of information that creates the most striking contrast between the expertise of these hypothetical individuals. Imagine the associations and methods which might be conjured up by our chef when considering how to cook meat with fruit, for example, and how these would differ from our occasional cook's take on the same subject. See also Chapter 15: "Using Existing Data" on the differences between experts and novices.

Passive memories can be episodic or subject matter related. The same is true for intentional memories. We can create a two-dimensional diagram of memory types. The diagram is a useful tool for understanding audiences' relationships to information.

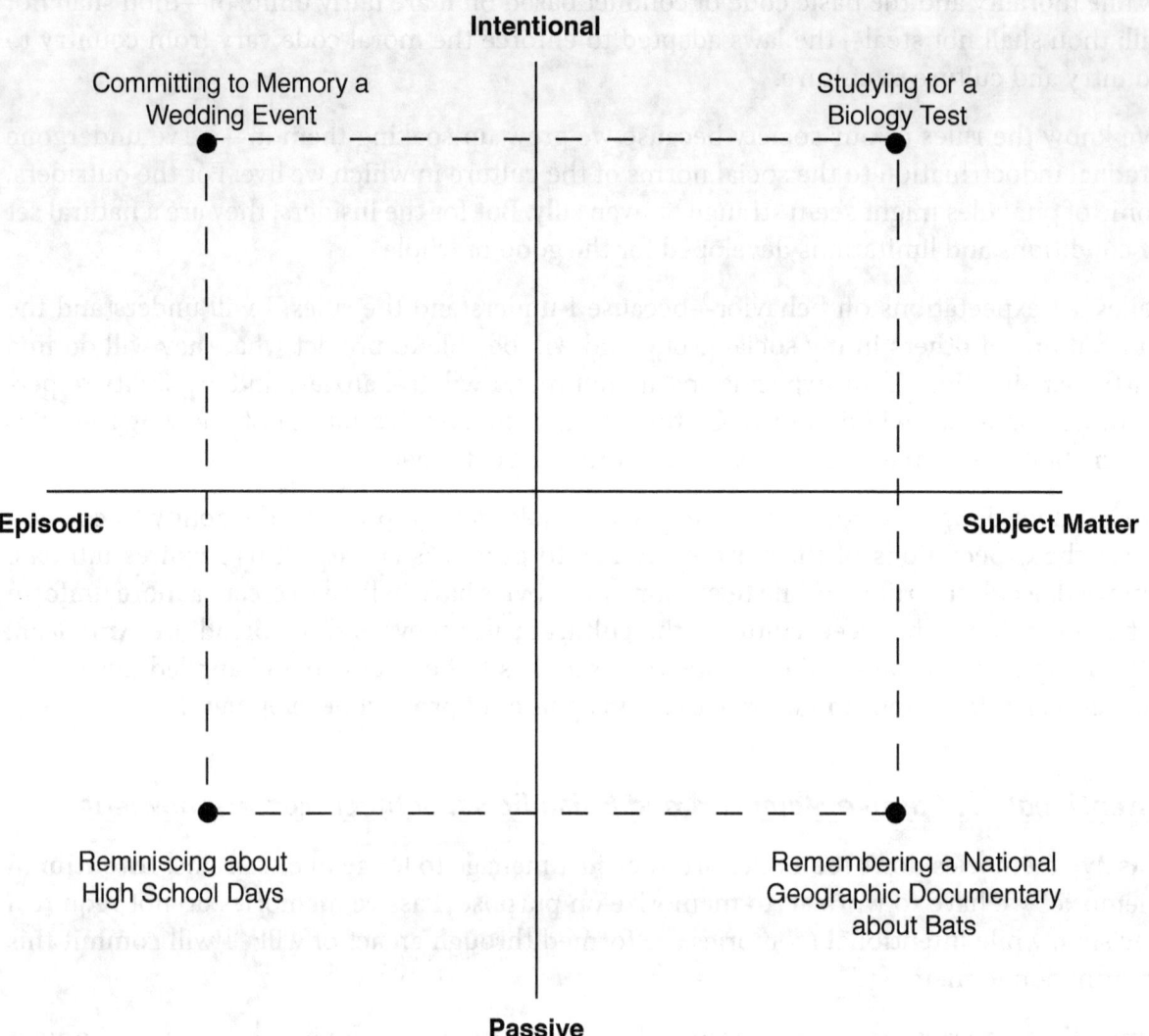

Level of Detail: Leveler vs. Sharpener

One should absorb the colour of life, but one should never remember its details. Details are always vulgar.

—Oscar Wilde

"Leveling" and "Sharpening" are the two poles of a continuum which define differences or variations in memory processing, specifically, the ability to retain discrete images of sequential, experience-based stimuli and to make distinctions between them. These stimuli may be verbal, as in a story, or visual, as in a film. **Levelers** are likely to over generalize events, objects, or ideas. **Sharpeners** tend to concentrate on the differences and rely on rote memory. In normal everyday situations, one's leveling/sharpening style remains fairly stable. However, it may vary substantially when confronted with an unusual or unexpected situation. Depending on the lack of control an individual feels during a highly stressful situation, he or she may move to either end of the continuum.

Leveling and **sharpening** describe an individual's memory manipulation style. It's similar to Narrow/Wide Categorization Style and Focuser/Scanner (perceptual processing characteristic) in its effects on human performance—but the root is different. Even so, it seems that it would be difficult for a particular individual to be a **leveler** when it comes to long term memory but act as a **focuser** when it comes to gathering information through the senses. For more information about the Narrow/Wide Categorization Style and the Focuser/Scanner characteristic see the relevant section in Chapter 10: "How We Think About Problems."

Additional Thoughts and Further Readings

You can only predict things after they've happened.

—Eugene Ionesco

Mel Levine's book, "A Mind at a Time," is a good resource for more information on memory, as is Anita Woolfolk's "Educational Psychology." Both those books were recommended at the end of the previous chapter.

What do you think these signs mean?

Designing traffic signs is a particularly interesting problem: they need to be memorable; they provide information to individuals with a rather taxed working memory; they have to "remind" the drivers of rules of the road; they need to be intuitive and as culture-independent as possible (a traffic accident caused by a tourist can still result in loss of life); they need to be seen, processed, and understood in seconds; and they can only occupy a very limited space.

A game of "Set" is a particularly good card game: very simple, yet fun to play. The task is to match three cards based on three characteristics: shape, color, and fill. The three cards form a set when they are either all the same or all different for each of these categories. This game relies on pattern recognition and memory. To learn more about this game, visit www.SetGame.com

9. Background Knowledge

Any system which is sufficiently complex not to be understood by the user will be capable of behaving in ways unpredictable to that user, and when it does then the user may attribute this wayward activity to inherent personality—as they may with other humans.

— J. & G. Underwood

Introduction

Everyone approaches a task with the baggage of prior knowledge whether based on personal experience or on information learned from books, schools, work, social interactions, and other life experiences. This background knowledge is highly individualized, but some assumptions about what a particular group of people knows can be made if a developer is given information about the audience's culture, profession, age, and other relevant factors. The more specific the audience, the easier it is to target both the material and the manner of its delivery.

An individual's expertise is composed of knowledge that is learned and developed over a lifetime. Expertise can be divided into **background knowledge**—knowledge of a particular domain—and **meta knowledge**—knowledge about knowledge or knowledge about how to acquire more knowledge. Background knowledge is further composed of **Formal Background Knowledge** and **Informal (Intuitive) Background Knowledge**.

Background Knowledge:

- **Formal Domain Knowledge**
- **Informal Domain Knowledge**
- **Meta Domain Knowledge**
- **Formal Social Knowledge**
- **Informal Social Knowledge**

Our informal background knowledge is not limited to scientific or mathematical concepts. It's all the stuff that we were exposed to during our lifetimes but which we never actually took the time to investigate or learn in some more structured way. It's the information that we heard from someone once as opposed to that which we read in a book on the subject, for example. These snippets of ideas can be accurate or completely false.

When it comes to interaction design, consider the audience as divisible into computer novices and computer experts in addition to domain knowledge novices and experts. Computer novices have to be supported not only in interacting with particular content, but also in the use of the computing environment.

If the interface designer is now a computer expert, it's instructive to change platforms to an unfamiliar one to recapture the experience of being a novice. We are all novices at certain tasks. Like an actor using "the method," it's good for a computer interface designer to remember back to their own past to recall what being a computer novice feels like.

Formal Background Knowledge

Formal Background Knowledge is well-articulated and well-structured information about a particular subject.

In recent years a lot of research has been done to understand the nature of expertise in a particular domain. Here is a quick summary of what researchers have found:

1. An expert's knowledge and understanding of a subject tends to be largely intuitive and inaccessible to direct reflection. If you ask an artist why his paintings are created a certain way, his answer will most likely be vague. He might use such words as composition and color scheme, but ultimately, he won't be able to easily answer why his paintings look good. Understanding a subject area is very different from the ability to explain it and to teach it to others. And because an expert's knowledge gets so internalized, its representation gets further and further removed from his consciousness.

2. Experts tend to do more pattern-matching than rule-following. An experienced chef approaches cooking a particular dish very differently from a novice. A chef might pair an ingredient with a particular cooking technique—i.e. veal shanks with braising—while a novice would more likely look up a recipe and carefully follow its directions when preparing a certain dish. Only with experience can a person develop a feel for preparing food and be able to move away from rule-following and towards the general principles that govern food preparation.

3. An expert's knowledge tends to be more qualitative than quantitative. Using the example of a chef again, an experienced cook prepares food by feel rather than numbers—"It needs more salt. It feels medium rare. The crust needs to brown more."

4. An expert's knowledge also tends to be very context and domain specific. One wouldn't trust a trusted physician to pick stocks, unless you also happen to know that the doctor is good with investments.

Formal Background Knowledge Acquisition: Zone of Proximal Development

Photograph of Lev Vygotsky from the original Russian edition (1934) of *Thought and Language.*

Common sense is very uncommon.

—Horace Greeley

A well-designed interface of a product should serve as a bridge between what users already know how to do and what users don't know how to do. That bridge exists within the **Zone of Proximal Development** ("ZPD").

ZPD is a concept introduced by a Russian psychologist, L. S. Vygotsky, in his theories on human knowledge acquisition shortly after the Russian Revolution. ZPD is the zone between what students know how to do on their own and what they can't accomplish even with the help of a great teacher. It's what a student can achieve with guided assistance. We now understand that all effective teaching efforts should be aimed at this zone. Efforts aimed beyond this zone, according to Vygotsky, will be ineffectual. I hope that if you read Chapter 4, you're starting to think this sounds a bit like flow.

An interface designed for a pre-reading child, for example, could have words as a navigational device, but it should have other visual cues to supplement the text. After a while, the child will learn to recognize the text without other cues.

A simple example is the "yes/no" buttons, common on children's games. One approach is to couple the written text with little animated heads that nod affirmatively or shake their head negatively as appropriate.

It is critical to understand the audience for which the program is designed. This allows you to

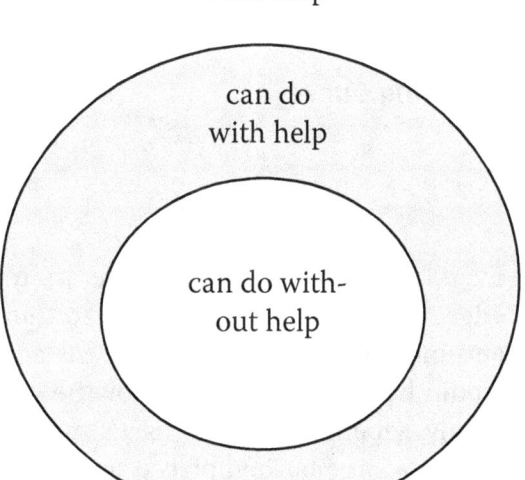

can't do even with help

can do with help

can do without help

determine their ZPD. Once you know, or have a good guess as to what a person is capable of doing with help, you can develop supporting structures for a particular task. These support structures are called "**scaffolds**" and they are a key concept to effective interaction design.

Zone of Maximum Benefit

At some point...we must have faith in the intelligence of the end user.

—Anonymous

It would be great to always be able to understand just what a user needs and to individualize the content to make those needs easier to meet. But in practice, it's usually just not feasible. There are always budgetary issues, there never is enough time, and it's hard to accommodate everyone all of the time. So design is all about compromise: what is the best solution within our means and doable within our time frame? How can we try to please the most number of users the maximum number of times?

The **Zone of Maximum Benefit** is our riff on the Zone of Proximal Development. While the design solution must lie within the Zone of Proximal Development, the execution lies within the Zone of Maximum Benefit—do the best with what you've got. In the context of teaching this material to fourth graders, we called the Zone of Maximum Benefit "ZOMBie." The kids talked about where their individual **ZOMBies** were.

Consider the example of curriculum development. It would be great to have individualized learning materials that tightly fit the cognitive strengths and weaknesses of each student. But that's just not possible, even with computer-based learning—there are simply too many variables. However, we can design and develop educational materials in a wide variety of formats: graphical organizers and text-based outlines; visual illustrations and mathematical formulas; audio books, paperbacks and movies; etc. The greater the variety of materials, the more likely some of them would fall within the preferred cognitive style of a particular student. This encapsulation of content can form the basis of scaffolded design support.

Learning Curve

Everything should be made as simple as possible, but not simpler.

—Albert Einstein

Even before taking into consideration users' cognitive differences and background knowledge, a product designer needs to think about the general characteristics of an interaction and interface that are desirable for a particular product. For example, what learning curve would be most appropriate for the project? The learning curve for using an application is highly dependent on the user's motivation to learn it. If the learning curve is high, there will be a large burden placed on the user to understand, memorize, and get proficient with

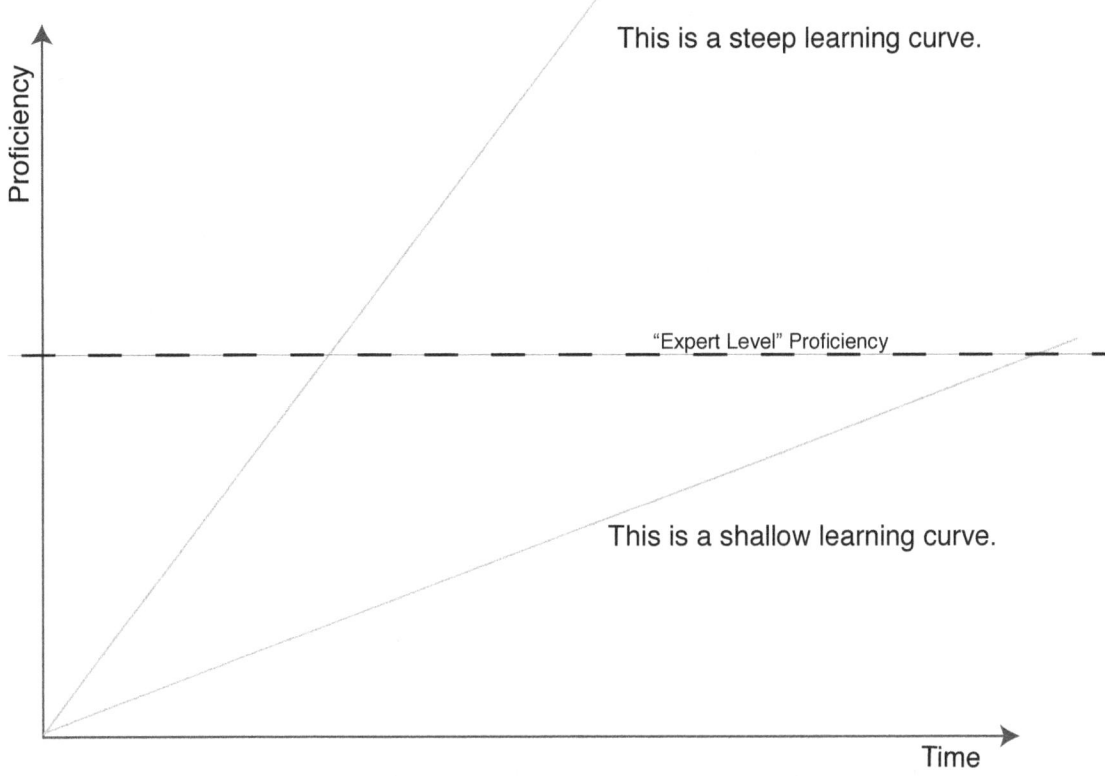

If the learning curve is high, there will be a large burden placed on the intended user to understand, memorize, and get proficient with the product. This means that the user is likely to make a lot of mistakes in the beginning and will have to be well motivated to invest the necessary time to learn to use the product.

the product before it becomes a useful tool. This means that the user is likely to make a lot of mistakes in the beginning just navigating and using the basic features of a system or application. A complicated system while offering more features might not be ideal for a casual user.

Consider an online learning example: you are interested in looking up a recipe for a sourdough bread an example of on demand learning. Would you be willing to invest your time in downloading, installing, and learning to use proprietary software to get it? That software could be great and allow such features as video-based cooking instruction, bulletin boards, real time chat rooms with professional bakers, video conferencing with people who are preparing the bread now, and so forth. But all you want is a paragraph worth of text with an ingredients list and some directions. However, if you were interested in taking an online course in maintaining nuclear submarines, the proprietary software might be just the ticket.

Some products are so complex that they demand a steep learning curve. That can be appropriate if the incentive of the user to master the software is greater than the difficulties of doing so. For example, a molecular modeling tool oriented to geneticists might need a huge number of features, require significant background knowledge and, appropriately, have a steep learning curve.

Even with a steep learning curve, there should be a built-in reward structure that both serves as an incentive and acts as emotional support for the user's hard work. With a shallow learning curve, conquering the product is a reward in itself. But for software with a complex learning curve, the user should see some results from their efforts right away. Adobe's Photoshop provides a good example. This software product is extremely complex, and even expert users often find new features and techniques. But even after a short investment of effort, a novice can get some results in Photoshop. As he learns more, he increases his capabilities with the program. So while Photoshop's learning curve is steep in total, mastering each feature is not that difficult and the user who does so feels rewarded.

Another characteristic of an interface is how memorable its features are over time. If the product is used every day, if it's part of a "tool box" of applications needed for daily survival, its features will soon be committed to memory. But if the product is to be used only once in a while, the features of its interface are likely to be forgotten or confused. Graphical user interfaces and integrated help options strive to solve the problem of retention and thus increase user satisfaction.

The user's emotional satisfaction while using a product is a critical characteristic of its interface. If users feel frustrated or stupid while using a product, the interface is a failure. The warm fuzzy feelings users get while interacting with a product are very important to the product's success.

Desired Level of Proficiency

An important feature of a learning curve is the **Desired Level of Proficiency**—how well do you expect users to know the product? There will always be errors; users will always make mistakes. What rate of errors is acceptable for the product? How catastrophic are the consequences of users' errors? If the user is an air traffic controller, his errors might result in an airplane crash and death. If the user is playing a game where he is an air traffic controller, the death and disasters are virtual and the failure is not catastrophic. These are important design considerations.

If the dress doesn't fit, most times the results are not life-changing.

Pattern of Use and Learning Curve

The desired level of proficiency depends on the pattern of use: how often the product is used, how important or how indispensable it is to the user, and how much automatization have the users achieved? Some products are used seldom enough that people tend to forget how to operate them in the intervals of time between use. Since the learning curve is all about achieving the desired level of performance, product developers have to factor in the pattern of use into the design. If the performance drops between uses, what can be done to quickly get users back up to speed? When the product has to be learned and relearned over and over again, its learning curve looks like a wave.

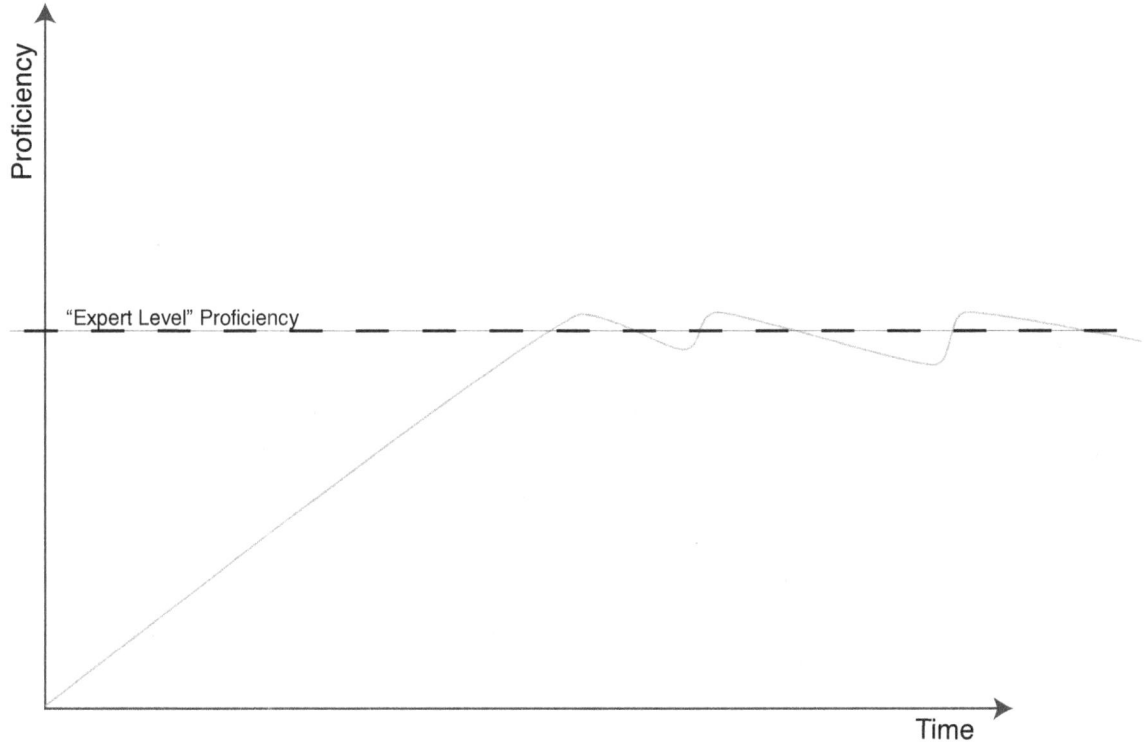

Note that each time it takes less time to relearn how to use the product; a user can achieve the desired proficiency level in less time. And it also takes longer to forget how to use the product with successive experiences with it; the slope of the learning curve between use is more gentle.

Learning Curve and Doctors

How doctors learn their skills is an open secret that we as a society don't like to talk about much. Clearly, to get good at a particular operation, surgeons have to have lots of practice on real people. Doctors learn through an apprenticeship model: they watch an expert do a procedure; then they assist; then do one themselves; then they do a lot of them; and finally they teach others. It's the cycle that drives medical education. When you stay at a teaching hospital, it's hard to dwell on the fact that the doctor assigned to insert a tube into your heart, could be doing this for the first time.

Informal Background Knowledge

We believe it is this tendency, which is a very strong one, to be impressed with the rare events and to ignore the common ones, that accounts for the fact that some geologists and water engineers give too much credit to the exploits of the diviner.

—Evon Z. Vogt and Ray Hyman

We showed a preview of the movie "The Passion of Christ" to our sons, eight and nine years of age at the time. Upon seeing these clips form the movie, our eight year old exclaimed: "Oh, Simone would love this movie." Simone was our babysitter. Simone might like the movie, but we wanted to know why our son thought so. He told us that Simone always says: "Jesus Christ!" Clearly, she would love the movie. This anecdote is an illustration of the application of informal background knowledge to a particular situation—for our son, his logic was irrefutable.

Just to drill the point a bit deeper, here is a question asked by our friends' six year old son Max: "When did the world turn color?" Consider his point of view: there were all those black and white TV shows and movies, and then... As a designer, you have to know your audience!

Informal Background Knowledge is that "hard to access," "not easily put into words," intuitive understanding of the world around us that we have been acquiring over our lifetimes. This informal knowledge can be deconstructed into social and phenomenological components. These components are triggered by social settings or situations. For example, one might be able to tell you that a year is 365 days long but not be able to tell you how long it takes for the Earth to make a complete circuit around the Sun. Our understanding of calendars is not always connected to our understanding of the mechanics of the solar system.

P-prims

The most overlooked advantage to owning a computer is that if they foul up there's no law against whacking them around a little.

—Porterfield

We all walk around with vast libraries of knowledge stored in our long term memories. Some stuff we know—we can state an idea and come up with good explanations for it. But some thoughts are slippery, hard to hold, difficult to put in context. These snippets of ideas provide the basis for a lot of things that we do in our lives, yet we don't have a good grasp on what they really mean, how they fit into other ideas we have, or even if they truly have any reality. These slippery snippets of ideas are **phenomenological primitives**, or **p-prims**, as defined by Andrea diSessa, Professor of Education and Cognition at the University of California at Berkeley. He describes them in the following way:

> "P-prims are relatively minimal abstractions of simple common phenomena. …[Novices] have a large collection of these in terms of which they see the world and to which they appeal as self-contained explanations for what they see."

A particular individual has a large collection of p-prims from which he or she can make causal inferences about a situation. Such a collection of p-prims does not have to be internally consistent—they depend on aspects of the problem-solving situation. Some explanations "come to mind" under one set of circumstances and other explanations are invoked by different situations. The same individual can generate contradictory explanations of the phenomenon under different circumstances.

These properties of p-prims can be summarized by two attributes:

- **Cuing Priority**—how likely the p-prim is to be profitable.
- **Reliability Priority**—how resistant the p-prim is to change.

If a particular p-prim is useful in generating explanations across many situations, then it has a high cuing priority—it "comes to mind" a lot. A p-prim that has served an individual well would be difficult to modify and would have a high reliability priority. "Water puts out fire" is a standard p-prim which works most of the time. Many people don't know why water puts out flames, but this doesn't stop them from using this information. Unfortunately, grease fire get worse when dowsed with water. This p-prim has cause a lot of injuries and even death—just ask any firefighter.

Proverbs can be considered p-prims in the folk wisdom category. "He who hesitates is lost" may contradict "Look before you leap," but both may be part of our mental constructs.

Examples of P-prims

> *The real problem is not whether machines think but whether men do.*
>
> —B. F. Skinner

Recently, I had to spend a few nights in the hospital. I was running a fever and the nurses prodded me every hour to take my temperature. The one thing that one hopes and expects from the medical staff at the hospital is to be able to take a reliable temperature reading. But over and over again, this illusion of competency was rudely broken. When my temperature reading registered well below normal, I suggested the nurse not use the ear thermometer and try the oral variety—it takes longer, but the results tend to be more accurate. The oral thermometer registered a high fever, a full three degrees variation in the readings from the ear thermometer. Upon seeing that I, in fact, was running a temperature, the nurse soothingly told me that sometimes it's just a lot hotter under my tongue then in my ear—I shouldn't worry too much. I was too sick to point out that thermometers are calibrated to insure accurate reading regardless of the place they are inserted. Clearly, the ear thermometer was malfunctioning (the nurse could have guessed that I had a high fever by the color of my face and by putting her hand on my forehead). But the nice nurse recorded the low temperature reading when I told her that I would like to get out of the hospital as soon as possible (can you blame me?)—a high fever would have kept me there for another day. This misunderstanding of the basic workings of the thermometer and over reliance on instruments is common, in my experience, among the nurses who take temperature readings.

P-prims are not limited to temperature readings, computer users have plenty of slippery thoughts when it comes to their computer interactions. Two examples of inappropriately cued computer-based p-prims were exhibited by sixth-grade science students using Macintosh computers: the "hide-a-file" and the "delete-a-file" p-prims. The "hide-a-file" p-prim is the false belief that the computer has the ability to relocate and hide a particular file amongst the other files. The probable origin of this p-prim is students' confusion over where they saved the file they were working on. Often, rather than ending up in their personal folders, these "missing" files end-up inside the folders containing the application with which the students were working. The "delete-a-file" p-prim is related to the "hide-a-file" p-prim. It refers to the belief that the computer has the ability to randomly delete files which students have created. Again, these files were often just misplaced. This sense of the computer as a diabolical agent is derived from the students' misunderstanding of its system of operation.

Misunderstanding of underlying causes is one of the origins for false p-prims. Another is the **misperception** of events leading to a particular outcome or misunderstanding of the causality of events. Imagine that every time you left the house wearing a red shirt, you got a speeding ticket. You might mistakenly believe that there's a causal connection between the two—the cop sees the color red and reacts to it badly. You decide that you'll just stop wearing red. Deep down inside, you might know that this is just a superstition, but many other similar connections form a core system of beliefs among some populations.

To test the slipperiness of thought, try asking someone to explain the origins of low and high tides. You will encounter a lot of hand waving before you find a person with solid understanding of the underlying physics. And unless you understand the phenomena yourself, it would be difficult to identify all of the p-prims coming your way.

Examples of Kids' Slippery Thoughts

Below are samples of kids' science writing gleamed from an anonymous email humor forwards.

"The moon is a planet just like the earth, only it is even deader."

"Planet: A body of Earth surrounded by sky."

"A fossil is an extinct animal. The older it is, the more extinct it is."

"Dew is formed on leaves when the sun shines down on them and makes them perspire."

"A super-saturated solution is one that holds more than it can hold."

"When you smell an odorless gas, it is probably carbon monoxide."

"The pistol of a flower is its only protection against insects."

"The skeleton is what is left after the insides have been taken out and the outsides have been taken off. The purpose of the skeleton is something to hitch meat to."

While these are funny, it is sobering to know that most adults believe the rise in temperature during the Summer months in the Northern Hemisphere is caused by the Earth passing closer to the Sun in its elliptical orbit.

Formal Knowledge of Social Behavior

Most men live like raisins in a cake of custom.

—Brand Blanshard

Formal knowledge of social behavior is the culture-specific knowledge of social interaction that comes in a codified and well-articulated form. Think, for example, of Ms. Manners or Emily Post. It is important to understand that what is socially acceptable in one culture is not necessarily okay in another.

Social Behavior in an Online Setting

> *As you make your bed, so you must lie in it.*
>
> —Daniel J. Boorstin

Most people in America have heard about computers and the Internet. And quite a few were formally introduced to one or both in a school, community, or a family setting. These introductions not only conveyed facts, procedures, and rules but also social attitudes and behaviors appropriate to computer settings and use.

The online world is evolving its own social conventions distinctly removed from those of the real world. Email is a good example. Because of the lack of affect conveyed in a plain email message, people have a great degree of difficulty trying to divine the emotional content of an email message. "Are they mad at me?" "Why are they being so cold?" The little iconic "emoticons"— ;-) or :-(—are an attempt to add emotional spin to a message. And most users know that it is considered rude (shouting) to send a letter in all capital characters. These conventions are evolving over time and are being passed on as social rules of behavior in an online setting among groups of users.

Scripts: Definition

> *Education is learning what you didn't even know you didn't know.*
>
> —Daniel J. Boorstin

A **script** is an archetypal story about a given situation and a set of expectations that include rules of behavior. People have an enormous collection of experiential scripts that they use everyday for the interpretation of events arising in many different situations. Problems can arise when scripts are applied inappropriately, or when an individual doesn't have a script to handle a particular situation.

It seems appropriate to use the same classification for scripts as for p-prims (see the previous). Thus individual scripts can be defined by two properties:

- **Cuing Priority**—how likely is a particular script to be profitable.
- **Reliability Priority**—how resistant is a particular script to change.

If a particular script is useful in generating explanations of behavior patterns across many social situations, then it has a high cuing priority—it "comes to mind" a lot. And a script that has served an individual well would have a high reliability priority and thus would be resistant to change.

Scripts: Research Background

Roger Schank of Northwestern University introduced the concept of scripts in the context of artificial intelligence research. (Note that the cuing and reliability priorities are concepts developed by Andrea diSessa and are applied here to scripts without the approval of either researcher.) Currently, Roger Schank is using the concept of scripts to develop an interactive language learning system that not only teaches basic vocabulary, but also the situations in which the vocabulary is appropriate:

> "Where is a bathroom?"
>
> "Where is the baggage claim area?"
>
> "Can I see a menu, please?"

The above phrases are put into the appropriate scripts of behavior and shown as video clips to the language learner.

A classic example is the restaurant script. One goes to a restaurant, expects to be seated and be given a menu, orders dishes, has the food brought by a waiter, receives a bill, pays the bill, and leaves a tip. The basic script allows those experienced with eating in restaurants generally to eat in a new restaurant and yet still share a set of expectations with the restaurant which guides their mutual interaction.

A simple example of a computer use script might be the manner in which applications are launched. Once someone has some experience with opening applications on a computer under a particular operating system, he develops a set of expectations about the procedure. If the user attempts to launch an application and those expectations are not met, he could become confused, lost, or presume that something went wrong—often feeling that he was to blame.

Scripts: Examples

Most of us visit the dentist on fairly regularly. And over the course of many visits, first with our parents and then by ourselves, we have developed the Dentist Script. Now imagine walking into the dentist office and the doctor asks you to eat an apple—he just wants to see your biting style. Then he asks you to do the same in front of the mirror—he wants you to be aware of your biting habits. The entire time you're having this exchange, there is a video camera focused on your face—the doctor wants to have a record of your teeth in action so he can compare it to movements made at a later date. All of this seems reasonable: why shouldn't your dentist be interested in your teeth in action? But at some point, you might feel that you've stumbled into some oral weirdo. While all of the above might objectively be reasonable actions for a dentist to take, they are not part of the normal "visit to the dentist" script.

There are also the "visit the doctor" scripts. We get into the office and sit on the examination table. We expect the doctor to put a stick into our mouth and asking us to say "ah."

The doctor is expected to listen to our chest and pound on it with his fingers. He should shine a light in our eyes and check our reflexes with a little rubber hammer. While all of the above are still valuable diagnostic tools, the medical profession changed. There is less time to do all this "hands on" stuff, and most young doctors are no longer even trained to understand the sounds they hear in our chests. So what's the point of listening if an x-ray would do a better job? For younger patients there might be little point—they have assimilated the new "visit the doctor" scripts, scripts that have been changed with the introduction of HMOs. But older patients still demand the old treatments and feel cheated when the doctor doesn't take the time to check them out "properly." These patients have the old "visit the doctor" scripts and their expectations are set by it.

Scripts and Culture

In both of the scripts described above, the consequences of diviating from them are not very drastic. You might change your dentist or feel shortchanged by your doctor, but nothing life threatening happens. But there are situations when the wrong script could get a person into serious trouble. Consider another medical example: "get a second opinion" script. Upon hearing bad news from your doctor, it's good to get a second opinion, just in case. The "get

a second opinion" script is culturally specific—Americans do it, but Russians do not. In fact questioning the doctor's decree was unheard of during the time I lived there as a child. Some of the Russian emigres find it very difficult to get a second medical opinion even after years of living in America and soaking up American culture. Sometimes not getting a second opinion is just fine, but sometimes it leads to medical errors. And these medical interaction scripts are not limited to second opinions. People from certain cultures just never question the recommendations of their doctors. This inability to take control over one's medical care can potentially lead to disaster.

Meta Knowledge: Definition

Once children learn how to learn, nothing is going to narrow their mind. The essence of teaching is to make learning contagious, to have one idea spark another.

—Marva Collins

Meta knowledge is knowledge about knowledge and knowledge acquisition. One of the goals of successful instruction is teach people how to learn. Meta knowledge includes the ability to monitor and make conscious decisions about what kinds of Reasoning Styles and Categorization Strategies are appropriate for tackling a particular problem. Meta knowledge also deals with understanding the use of Language and how to harness Attention Controls. And while a learner might know how to learn and have strategies that help with learning new information, that same learner might not be able to explain what it is that they know.

Additional Thoughts and Further Readings

A little sincerity is a dangerous thing, and a great deal of it is absolutely fatal.

—Oscar Wilde

Ms. Manners was a syndicated column by Jean Abrahamson which answered questions about social etiquette. Individual stories have been collected into several books. "Manners Are Magic: 'You'll Thank Me for Telling You' Lessons on Life from Ms. Manners" by Jean Abrahamson and Adam Greber is in print. This is a fun book to read and might even surprise you.

Roger Schank of Northwestern University introduced the concept of scripts in the context of artificial intelligence research. Scripts are rules of behavior and expectations that an individual has for the interpretation of events arising in a given situation. In his

1990 book "Tell Me a Story: A New Look at Real and Artificial Memory," Roger lays out his ideas and provides multiple examples of how scripts influence the outcomes of interactions.

A few years ago, The New Yorker published an article by Dr. Atul Gawande, "The Learning Curve." It's an insightful article about medical education in America. Like no other piece of writing, it stresses the importance of good design when in comes to medical equipment. Improved design of the equipment used by anesthesiologists has greatly reduced accidental deaths during surgery. This article has been reprinted in the Oliver Sacks' edited collection of "The Best American Science Writing 2003."

For an interesting insight into forgetting, search Google for Hermann Ebbinghaus, Piotr Wozniak, and the "spacing effect." The spacing effect describes the pattern of forgetting newly learned information and it looks similar to the second learning curve diagram presented in this chapter.

Lev Vygotsky's "Thought and Language" was first printed in Russia in 1934. The English translation only became available a few decades ago: "Mind in Society." His work was revolutionary, no pun intended, and is still widely taught in psychology and education departments in US colleges.

10. How We Think About Problems

Expecting all children the same age to learn from the same materials is like expecting all children the same age to wear the same size clothing.

—Madeline Hunter

Reasoning Styles

Reasoning Style is a preferred sequence of cognitive events or actions that we rely on in any given situation. It's the way we think; the way we approach problem solving; the way we occupy our working memory.

Top-down vs. Bottom-up Reasoning Style

Different people process information differently. The **top-down/bottom-up** cognitive style is a bipolar information-processing measure that describes the way that individuals select and represent information. You're either one or the other depending on a particular situation—environmental conditions change the way we think! But most of us rely on "tried and true" methods of thinking that have worked for us in the past, and so people tend to stick with either one style of reasoning or the other.

Some have a **top-down** or **deductive** approach to solving problems—they see the

Reasoning Styles:

- **Top-down or Deductive**
- **Bottom-up or Inductive**
- **Story-telling or Abstractive**
- **Reflective**
- **Impulsive**

task as a whole and are able to approach it from multiple angles. They use a global, thematic approach to learning by concentrating first on building broad descriptions. Top-down individuals typically focus on several aspects of the subject at the same time and have many goals: working topics that span various levels of the hierarchical structure. These individuals prefer to reason from the general to the specific. They tend to form hypotheses first. Then they seek examples and information to confirm their theories. Information architects tend to think and work this way.

Others are **bottom-up** or **inductive** processors of information. When they are given a task, they prefer to approach it sequentially, one step at a time. They concentrate on details and procedures before generating an over-all scheme for information. They typically combine information in a linear sequence, focusing on small chunks of information that are low in the hierarchical structure, and working from bottom up. These individuals prefer to reason from the specific to the general. They tend to start with examples and then form hypotheses to fit those examples. If you are a detective and looking to solve a murder mystery, this is the style of reasoning you have to adapt by necessity. Thus certain professions require us to think a certain way. Content architects tend to think and work this way.

When designing an interactional environment, one should consider these different ways people approach problem solving. There might be a way of accommodating both these styles.

Abstractive or Story-Telling Reasoning Style

Individuals with a **story-telling reasoning style** solve problems and form concepts by relating all new information to their personal lives. They turn all new ideas into stories to which they can relate. These individuals immerse themselves in experiences rather than cogitate about them. They like being part of a group and tend to communicate well. This reasoning approach is somewhat similar to the "Type I Learning Style" or "Accommodative Learners" as defined by David Kolb (see Appendix). Sometimes, it's hard to tell the difference between a personality trait and a reasoning style—everything is connected to everything else.

Reflective vs. Impulsive Reasoning Style

Reflective individuals make more reasoned decisions than **impulsive** individuals. Impulsive individuals tend to jump to premature conclusions. Reflective people tend to excel at the recall of structured information, interpretation and comprehension of text, problem-solving, decision-making, setting their own goals, concentrating on information, comparing new knowledge with existing knowledge or beliefs, and metacognitive strategies.

This reasoning style clearly has a lot to do with attention controls. An individual with poor attention controls would tend to jump to conclusion without reasoning though all of the problem's variables and implications. This is yet another example how all the cognitive characteristics described in the Cognitive Wheel are all interconnected.

Categorization Strategies

Only the educated are free.

—Epictetus

Without making categories and taxonomies out of the items in the world around us, it would be very difficult to remember or learn anything. But this skill takes time to develop. The older we get, the more experience we have with a particular subject matter, the more useful are our taxonomies and the better we are at making them. Kids typically have more difficulty with making categorizations than adults.

And just like kids, adult novices have trouble forming categories and sorting ideas in a subject area with which they are not familiar. To sort concepts, it is necessary to know something about them. People who know very little about computers or the Internet, tend to make mistakes based on miss categorization. My parents often refer to our Internet and their Internet—to them, it's not the same thing. Although they have a speedy cable modem, we seem to them to have infinitely more access to information that they do, like we've got the deluxe version. They also frequently criticize the Internet for its stubborn refusal to give them what they want, as if the Internet was controlled by a single miserly corporate owner.

Another common confusion is a mix-up between an application and the information it displays. If "Google" is thought to belong to Internet Explorer, for example, then Firefox is not going to be an acceptable alternative.

Categorization Strategies:

- **Relational**
- **Analytic**
- **Differentiational**
- **Sequential**
- **Narrow**
- **Wide**
- **Appraisal**

Relational Categorization Strategy

Given a group of objects, individuals who use a **Relational Categorization Strategy** to sort objects and form concepts place objects into categories based on functional relationships between or among the objects. Relationals tend to use global or thematic classifications (e.g. "pirates,") rather than categorizing objects by singular aspects or details (e.g. "people with hats."). They look for underlying meaning in their categories versus the surface features of the objects.

Analytic Categorization Strategy

Given a group of objects, individuals who use an **Analytic Categorization Strategy** to sort objects and form concepts will focus on the details of the object rather than on the object as a whole.

Differentiational Categorization Strategy

Differentiational Categorization Strategy implies that these individuals prefer to reason and create inferences by comparing and contrasting ideas based on selected characteristics. These individuals are looking for differences between bits of information.

Sequential Categorization Strategy

Sequential individuals tend to use a linear, step-by-step organizing scheme as opposed to hopping around through the material. Sequential individuals are more comfortable with highly structured and methodical content. This is clearly related to the Sequential/Judging/Scheduling personality type, discussed in Chapter 11: "Personality."

Narrow vs. Wide Categorization Strategy

If asked to categorize a set of items, **Narrow Categorizers** would tend to make more narrowly-defined categories into which to place the items than would **Wide Categorizers**. Narrow categorizers are typically conservative, excluding items from groups, and forming narrow classes. When measuring cognitive categories, narrow categorizers prefer to exclude possibly inappropriate instances through over discrimination and limitation of the scope of their category ranges. Wide categorizers prefer to risk the inclusion of possibly inappropriate instances by expanding the scope of their category ranges and broadly generalizing. If you could to pick, you would want to be somewhere in the middle—not too wide and not too narrow.

Clinical social workers tend to be narrow categorizers by trade—it's best to view each case as an individual with a unique set of problems, although recognizing similar issues across all cases is important, too. Researchers, on the other hand, tend to be wide categorizers—they need to form wide groups that underplay differences as part of their work. Think, for example, of the statement: "Obese people tend to have a higher incidence of stroke and heart attacks."

Appraisal Categorization Strategy

While some of us belong to the "tried and true" tribe, others are of the "see what works best" alliance. These latter individuals use all reasoning strategies without preference. They are able to appraise the situation and choose accordingly. These individuals also score somewhere in the middle of the narrow categorization strategy versus wide categorization strategy scale.

We are not born with a set of categorization strategies, we learn them over time. Sometimes, our teachers explicitly show a specific method or taxonomy, other times we have divined the winning strategy on our own. The more practice we have classifying information, the more flexible we become at it.

One of the most important tasks an information architect has to perform is to make her organizational scheme visible and understandable to its intended user group. In part this involves figuring out how people tend to organize information in that subject area in general. Different subjects often have different preexisting taxonomies and different way to view and examine information.

Additional Thoughts and Further Readings

There are several illustration styles in this book. And these graphical elements all depict different objects. How can they be grouped? Organized? Labeled? Is there a natural taxonomy of graphical objects that are used in this book? Here are a few samples you have seen so far. What's the best organization scheme? Does the goal of the organization matter? Which strategy did you use?

One way to arrange the graphics in this book is by the subject matter they illustrate. Another is by type: diagrams, pen and ink drawings, photographs, cartoons, book covers. But clearly, the illustrations are created in different styles, so perhaps grouping all of the cartoons together into one taxonomic pile is a mistake. Should the authors of these art works be considered? Perhaps an alphabetical arrangement by creators' last name is the way to go. But if you are concerned about production, maybe you'd group by vector graphic versus bitmap.

One of the learning activities for kids in kindergarten is sorting. All kids are required to bring two or more objects to class. The whole class of five-year-olds sits around in a circle with the loot in the middle. Students take turns to sort the objects in particular groups: one kid makes piles, the others have to guess the sorting criteria.

11. Personality

We suffer primarily not from our vices or our weaknesses, but from our illusions. We are haunted, not by reality, but by those images we have put in their place.

—Daniel J. Boorstin

Introduction and Research Background

How we view the world and how we interact with it are partly determined by our personalities. Some of us feel great at any party and are able to easily communicate with strangers. Others view an invitation to a cocktail party as a hostile act, a kind of personal hell. Some of us are leaders, and others prefer to follow. Some of us pay more attention to our feelings, while others prefer to listen to their own thoughts.

Each person is unique. But we will try to find the similarities between people while grouping them into subcategories that are chosen to emphasize particular differences in personality traits. The degree to which we fall into one category or another is, of course, different for different individuals, but there are some things that we can deduce from examining a psychological profile of a particular person. For example, if a person is classified as mostly an introvert, then it will be harder to interest him into coming to an all night dance and scream party...but perhaps in costume? It might also be difficult to get him to become an active member on an online bulletin board.

If we, the product designers, can anticipate the psychological makeup of the expected audience during the conceptual phase of product design, then we'll have a higher chance of this product succeeding with that audience. Often, we need to design for a wide variety of personality types. But knowing the categories allows us to fashion different interactions for each, ideally working together to satisfy all.

Unfortunately, a review of the educational and psychological literature reveals a lot of confusion and overlap in the terms that define personality types. I've attempted to synthesize these definitions to come up with descriptions and terms that would be the most helpful to a product designer.

Expressive vs. Reserved Personalities

Reserved describes a person characterized by an internal cognitive focus or by a concern with their own feelings and thoughts. These people tend to be shy and don't function well in social situations. By contrast, **expressive** individuals direct their attention to the outside world—they focus on other people and social interactions. Reserved individuals would find this behavior uncomfortable.

In the United States, we tend to reward expressive personalities over reserved ones. But in Japan, if you are too expressive, you can be construed as rude. So while in America, 75% of population have expressive personalities, this percentage will be different for other cultures who reward this personality trait differently.

Emotional vs. Reasoning Personalities

Emotional people are said to follow their heart, which means that much of what they do is based on their emotions or desires. For them, it's not the content of the message, but its delivery. **Reasoning** individuals are said to govern themselves with their head, their concepts and precepts are their guides to action. Everybody has both feelings and thoughts but working memory limitations mean that it is difficult to pay attention to both at the same time. An individual cannot be simultaneously high on both the Emotional and the Reasoning scale. A person pays attention more to one or to the other at any

given instant. For further discussion, see Chapter 6: "Memory Metaphors."

Emotional and reasoning personalities are about equally distributed in the American population.

Schedulers vs. Probers Personalities

This cognitive style describes a difference in temperament, and thus is personality related.

Schedulers like to plan in advance. **Probers** like to take risks and are comfortable with uncertainty. Schedulers tend to be judicious, while probers are open to options. Schedulers make agendas, timetables, programs, lists, syllabi, calendars, outlines, registers, and so on, for themselves and others to follow. Probers keep their eyes open for chances to do things they want to and for opportunities and alternatives with which to avail themselves.

Schedulers and probers are about equally distributed in the American population.

Observant vs. Introspective Personalities and Working Memory

Observant people pay attention to the information which arrives through their physical senses, directly observing surrounding tangible reality. **Introspective** individuals prefer to direct their attention inward, using their senses to gain impressions which are then amplified by their imaginations. Introspective individuals tend to do a great deal of unconscious background synthesis; they pay attention to their thoughts or an internal dialogue. Observant people have more working memory allocated to observation—sensory information. For further discussion, see Chapter 6: "Memory Metaphors."

There are three times as many observant individuals than introspective ones in the American population.

Four Basic Personality Subtypes

Each person can be assigned a personality type based on the four variables described above. But out of all the variations, there are four basic personality types that show the widest and clearest differences in temperament. They are Observant and Probing, Observant and Scheduling, Introspective and Emotional, and Introspective and Reasoning. The full personality subtypes are refinements on these basic four.

It is important that these descriptions be viewed as only guides to behavior (although they have surprising predictive power). In a situation where little is known about prospective users, just knowing that they work for the customer service department of a large computer-based company, for example, can give the designer surprising clues to the daily habits and preferences of this group.

Observant and Probing Personality Description

When you think of a person with **Observing** and **Probing** personality think Bart Simpson.

People with this personality can be described by the following phrases:

- adaptable, artistic, and athletic
- aware of reality and never fighting it
- open-minded and on the lookout for workable compromises
- knowing what's going on and able to see the needs of the moment
- interested in storing up useful facts and having no use for theories
- easygoing, tolerant, unprejudiced, and persuasive
- gifted with machines and tools and interested in firsthand experiences
- sensitive to color, line, and texture
- generally enjoying life

Individuals with an observing and probing personality make up about 38% of the general population. For individuals with this personality type, it's all about the process, the end result is less interesting. And the process needs to be enjoyed—an individual with this personality will eventually (sooner rather than later) abandon an activity if he is not having a good time. If the activity is too repetitive or has little variation, it is not a good activity for this group. Thus drill and practice will not hold these people's attention for long. Lectures and presentations should be short, and so should reading assignments—this group works and learns in small-sized chunks.

This is a group that jumps from one activity to the next and is impulsive. The content has to grab a person and hold his interest in order for them to succeed. Games and simulations work really well. And these individuals thrive in problem-solving situations that challenge their skills. They are willing to take risks to succeed. They love hands-on activities and multimedia presentations. And they are very good at negotiating.

If bored, a person with an observing and probing personality is likely to act out, poking others, banging on furniture, and otherwise being disruptive. These individuals crave excitement and fun, even at work. They can be annoying to others around them, as they start far more projects than they ever complete.

From the data, this group is the least represented in universities and colleges and has the lowest correlation between ability and school performance.

The four observing and probing personality subtypes are:

Observant, Probing, Emotional, Expressive: Entertainer, about 13% of population

Observant, Probing, Reasoning, Expressive: Promoter, about 13% of population

Observant, Probing, Reasoning, Reserved: Artisan, about 5% of population

Observant, Probing, Emotional, Reserved: Artist, about 5% of population

Observant and Scheduling Personality Description

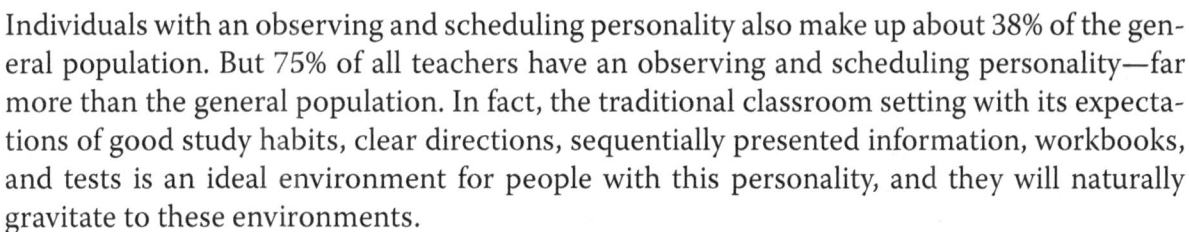

These phrases tend to describe a person with an **Observing** and **Scheduling** personality:

- conservative and stable
- consistent and routinized
- sensible, factual, and not impulsive
- patient, dependable, and hard-working
- detailed, painstaking, persevering, and thorough

Individuals with an observing and scheduling personality also make up about 38% of the general population. But 75% of all teachers have an observing and scheduling personality—far more than the general population. In fact, the traditional classroom setting with its expectations of good study habits, clear directions, sequentially presented information, workbooks, and tests is an ideal environment for people with this personality, and they will naturally gravitate to these environments.

People with an observing and scheduling personality are obedient and conform to standards set down by authority. They thrive on stability and are likely to generate procedures for doing things. Independent projects would be uncomfortable for these individuals who prefer question and answer sessions and the Socratic method of learning. This group tends to be extremely responsible: observing and scheduling people tend to do the right thing at the right time.

The four observing and scheduling personality subtypes are:

Observant, Scheduling, Emotional, Expressive: Seller, about 13% of population

Observant, Scheduling, Reasoning, Expressive: Administrator, about 13% of population

Observant, Scheduling, Reasoning, Reserved: Trustee, about 6% of population

Observant, Scheduling, Emotional, Reserved: Conservator, about 6% of population

Introspective and Emotional Personality Description

These phrases tend to describe a person with an **Introspective** and **Emotional** personality:

- humane and sympathetic
- enthusiastic and religious
- creative and intuitive
- insightful and subjective

Individuals with an introspective and emotional personality make up about 12% of the general population. To succeed, such individuals need well-established routines and well-defined expectations and goals. They do better with personalized attention—customization will be important to them. These people work well in groups and in democratically-run environments. They enjoy and thrive in social settings. They define themselves in terms of their relationships with others. People with an introspective and emotional personality are also more likely to participate in polls and share their ideas with others in their community. These individuals tend to have strong communication skills and particularly excel at writing. On the Internet, these surfers gravitate towards chat rooms and BBS's, and they are more likely to use instant messaging as a continuous form of communication with their peers. They tend to be pleasant and agreeable. These individuals are more likely to gloss over the details and are happy with general impressions and the "big picture."

The four introspective and emotional personality subtypes are:

Introspective, Emotional, Expressive, Probing: Journalist, about 5% of population

Introspective, Emotional, Reserved, Probing: Questor, about 1% of population

Introspective, Emotional, Expressive, Scheduling: Pedagogue, about 5% of population

Introspective, Emotional, Reserved, Scheduling: Author, about 1% of population

Introspective and Reasoning Personality Description

These phrases tend to describe a person with an **Introspective** and **Reasoning** personality:

- analytical and systematic
- abstract, theoretical, and intellectual
- complex, competent, and inventive
- efficient, exacting, and independent
- logical and technical
- curious, scientific, and research-oriented

Only about 12% of the general population falls into this group. Scientists tend to predominately have introspective and reasoning personality. People with this personality are independent learners and tend to have a strong interest in technology. They like tracking down information. They are interested in logic and respond well to reason. These individuals have a strong sense of justice. They like to keep control and are fairly self-sufficient. They also need constant success to continuously prove competence. Individuals with this personality want to be recognized and valued based on the quality of their ideas. And while they are strong on reasoning, they tend not to be good at communication—thus activities which involve written communication are not optimal for this group. Introspective and reasoning people don't like to bother with redundant or useless paperwork—this characteristic might easily translate into aversion towards the needless busy work so often performed in the classroom or in an office.

The four introspective and reasoning personality subtypes are:

Introspective, Reasoning, Expressive, Probing: Inventor, about 5% of population

Introspective, Reasoning, Reserved, Probing: Architect, about 1% of population

Introspective, Reasoning, Expressive, Scheduling: Scientist, about 1% of population

Introspective, Reasoning, Reserved, Scheduling: Field Marshal, about 5% of population

Anxious Personality

Anxious people are worriers, and most of their working memory is taken up by processing their feelings about the situation rather than analyzing the problem at hand. These individuals don't deal well with uncertainty and require a lot of emotional support. They also require a great deal of structure when they first encounter a new instructor, a new material, or a new situation.

The amount of concentration one can bring to bear on a task depends, in part, on the number and quality of the distractions. Environment affects the working memory—anxiety and stress about a particular situation affects individual performance. If a person is spending most of her working memory on worrying, then she has less working memory capacity to dedicate to the task itself. This is why some students are not good at taking tests—they might know the material, but the stress of the test situation diminishes their mental resources.

Collaboration, Cooperation, and Competition

The interactions between members of a group can be classified into competitive, cooperative, and collaborative activities. The basic definitions of both collaboration and cooperation are two or more individuals working together on a project. But there is a lot of freedom in this definition. To be useful instruments in examining group dynamics in a particular setting, differences between collaborative and cooperative interactions among group members need to be specified.

In common speech, cooperation and collaboration have multiple meanings, and are even sometimes used interchangeably. But they don't mean the same thing. They express different types of interactions among participants in a group. When a hundred Russian peasants dragged the barges up the river, they worked together, collaborating on every pull of the rope. But when a crew of sailors run a ship, they each have a distinct job to do, even though their goal is the same. Do these sailors have the same degree of collaboration in their work as the peasants pulling the barge?

In a **cooperative interaction**, the overall goals are shared by all of the participants of a group, but the work load can be distributed in many different ways. Cooperative group members can make different contributions to the whole: some might take charge of the project's scheduling, some produce graphics and/or written materials, and others provide data. If group members work on different parts of the project, it's important to analyze the individual contributions and responsibilities to the whole. The relevant questions are: "Are all participants equally responsible for the overall project?" and "Are there disparities in work loads?" In a **cooperative** task, the work load does not have to be distributed equally among the group and usually is not. **Collaboration**, on the other hand, specifies that group members work together on all aspects of the overall project: all contribute to writing, managing, data collecting, and so on.

Making this distinction between collaboration and cooperation is helpful to make clear the expectations upon individual contributors to a group project.

Collaboration, cooperation, and competition are explored here from the point of view of an individual working on a particular project. In collaborations, everyone knows each other and each other's capabilities, groups tend to be small, and projects limited in scope. In cooperations, some group members know each other and some don't. Cooperative groups can be large with members having limited knowledge of what others are doing, and group members can come and go during the project's tenure. But individuals in both cooperative and collaborative projects share goals for the overall project and contribute their work towards achieving those goals.

Competitions are different in this respect. Competitors don't share any of the workload among themselves. They might not know the individuals that are competing against them or what they are working on. There is no shared information or work. The individual competitor's goal is to beat the others.

So how do these different group dynamics influence how people with differing personality types feel about working together? **Collaborative** individuals prefer to work with others in a group, and they do it well. Others enjoy working with them in a group setting. Collaboratives are good at making contributions to the whole.

It's important to consider what it means to have a collaborative situation. When two people work together, they could be collaborating or just working side by side. So just having more than one person working on a project is not enough to set up a collaboration.

When there are a hundred people working together, are they collaborating more than two individuals working on a project? Does the number of people involved change to degree of collaboration?

Sometimes, people work together on the same project at the same time. And then there are collaborative situations which are asynchronous. Does the quality of collaboration change when people don't work on the project at the same time?

As mentioned above, collaboration can mean a lot of things to a lot of people. So when describing a user group as collaborative, make sure that you understand the "flavor" of collaboration these people are used to—they might not be comfortable with other varieties.

Here are a few variables to consider:

- **Duration**—Is it a one-time thing or a continuous collaboration? Is it short-term or long-term working arrangement?

- **Number of Participants**—Are there a lot of individuals involved or few?

- **Degree of Required Participation**—Do all collaborators contribute the same amount to the job?

- **Volition**—Are the participants collaborating out of self-interest or out of necessity? Who generates the necessity?

- **Constitution of the Group over Time**—Are the same people collaborating for the duration of the project or are individuals free to come and go?

- **Reward**—Is the collaboration set up as a self-rewarding experience or is there an outside incentive (e.g. prize)?

Depending on the values of these variables, the structure of the product and its usability will have to be conceptualized differently. Some collaborative structures are really disguised ways to foster competition and as such disappoint the expectations of users seeking a collaborative experience. For an example of Web cooperation, read about The Company Therapist project in Chapter 19: "Design Recommendations."

Competitive individuals are more motivated by competition with others. They enjoy competition and they tend to want to boost their self-esteem through besting others. They tend to measure themselves against the accomplishments of others. These people tend to brag and exaggerate their own accomplishments.

Obsessive Personality

Obsessiveness is defined as unwanted, unpleasant, uncontrollable, repetitive, recurrent, and persistent actions and thoughts. Individuals with obsessive personalities have difficulties with transitions from one activity to another. Such people can develop rituals which they feel compelled to perform.

Purposeful Destructive & Disruptive Behavior

Destructive behavior can be self-directed or can target other individuals. It is composed of socially unacceptable actions. Self-destructive behavior includes self-mutilation and the infliction of physical pain, destruction of personal property, and taking recreational drugs. Purposeful destructive behavior targeting other individuals includes bullying (verbal and physical), destruction of property, stealing, and arson, for example.

Disruptive behavior is characterized by its effect on others around the person: actions that result in disruption of group work, class work, team work, family life, interactions with co-workers and friends. Disruptive individuals have a hard time following rules and deliberately break them.

For product designers interested in constructing social interactions, the ability to foresee and diffuse the actions of destructive and disruptive individuals is important. These users can cause product failure in situations demanding social interaction. Some online games, for example, are plagued by aggressive players who like to immediately dispatch newbies as they first materialize into the game. When foreseen, such disruptive behaviors can be curtailed through interaction design.

Attitude and Intelligence

Attitude describes a person's feeling towards work, coworkers, school, friends, and self in relation to others around him. These feelings, like other thoughts, occupy space in working memory and make it difficult to focus on other activities. Negative feelings towards a particular person or a particular activity can interfere with an individual's performance. Similarly, negative feeling towards one's personal abilities can lead to lowered standards of performance, feelings of hopelessness, and even depression.

There are two ways of thinking about intelligence. There's the **static theory** of intelligence—you are born with a certain amount of brain power and you have to learn to live with it. The contrasting theory of intelligence is **incremental growth (dynamic theory)**—the more you learn, the more you know, and the more intelligent you become.

Research has shown that people who believe in a static theory of intelligence do worse in school than those who believe intelligence grows over lifetime. The reason is attitude. Like the length of your fingers or adult height, if you believe in the static theory of intelligence, you believe that there's not a lot that you can do about how much intelligence you were endowed with at birth. If you believe that no amount of learning can change how intelligent you are, then every test you take and every project you undertake is a test of your intelligence. And every time you fail is yet more proof to the world that you are dumb. On the other hand, if you believe that your intellectual abilities grow with education and experience, then failing a test just means that you didn't study. Such failure makes no absolute pronouncement about who you are.

An individual's personal theory of intelligence doesn't only impact his own performance, it colors how he views the abilities of others.' Employers who believe workers are not likely to improve due to lack of intelligence will act accordingly. Most don't even realize that they might treat employees differently based on how they perceived the employees' intelligence. If employers have low expectations of their workers, the workers will feel the same about their abilities. The same is true of friends—if friends don't believe their buddies are capable of more, their buddies will rise to the level of their non-expectation.

Computer Affinity

If computers get too powerful, we can organize them into a committee. That will do them in.

—Bradley's Bromide

Individuals with high computer affinity enjoy working with computers and feel comfortable with computer-based tasks. Individuals with low computer affinity would rather be involved in non-computer related tasks and experience stress in a computer-based environment.

This is an emotional response to the use of technology. As such, it should probably be part of the Anxious Personality classification. It is also related to Background Knowledge—anxious individuals tend to be more anxious around things they don't know or understand.

Additional Thoughts and Further Readings

Personality differences and Learning Styles have been used to sell specific product design concepts. Appendix of this book contains a brief literature review to help you differentiate key ideas and identify the players. The main thing to understand and remember is that Learning Styles and Personalities are really describing the same set of concepts and characteristics.

In 1984, David Keirsey and Marilyn Bates wrote: "Please Understand Me: Character & Temperament Types." David followed this volume up with "Please Understand Me II." Both books contain complete descriptions of personalities and a few tests. David also runs a Web site that provides some of this information online: http://keirsey.com. Unfortunately, the site now charges money for more complete access, but there are still fun ideas left to explore for free.

To examine cooperation and collaboration interactions in greater detail, read Dillenbourg's exploration of collaborative learning in his 1999 book "Collaborative-Learning: Cognitive and Computational Approaches." Or you can read my article for Ed-Media 2007: "Examination of Student Motivation and Group Dynamics in the Internet-based Learning Experiences," free for download off my company's Web site: www.pipsqueak.com.

For more information about different views of intelligence and other factors influencing school performance, read the 2000s edition of "How People Learn: Brain, Mind, Experience, and School" by Bransford et al.

12. Perception

It is in fact nothing short of a miracle that the modern methods of instruction have not yet entirely strangled the holy curiosity of inquiry. … It is a very grave mistake to think that the enjoyment of seeing and searching can be promoted by means of coercion and a sense of duty.

— Albert Einstein

In the book "The Island of the Colorblind," Dr. Oliver Sacks describes his adventures through a series of islands in Micronesia, which have an unusually high percentage of congenital achromatopsia (sever colorblindness) among the population. One of his traveling companions was a Norwegian scientist, Knut Nordby, an achromatope himself. Dr. Nordby described how his rather rare condition affected his early education. In addition to complete color blindness, congenital achromatopsia causes extreme sensitivity to bright light and poor visual acuity. Dr. Nordby's condition was diagnosed early, and since his vision was thought too poor to learn to read, he was sent to the school for the blind. But he hated Braille and used his sight to decipher the shadows on the page made by the raised bumps. He "cheated" at Braille and was caught. His teachers punished him for using his eyes to read and made him wear a blindfold in class. Dr. Nordby ran away from that school and eventually taught himself to read printed words and became a successful psychologist.

Dr. Nordby's story seems to be a cautionary tale of zealous educators trying to help a child with perceptual difficulties. Yet perception plays an important role in learning and information gathering—to understand, a person needs to have access to information, whether it is visual, auditory, or haptic. I use the term haptic, in this context, to group all information coming from senses other than sight and hearing.

When it comes to perception, we are all very different. Just in my immediate family of myself, my husband, and two children, we have two hearing impaired, one color blind, one smell impaired, and three in need of glasses. Fortunately for us, these particular issues don't carry a lot of stigma. Once people know about a difficulty, most try to accommodate. Upon meeting new people or starting a new class in school, my son makes a public announcement of his hearing problems. Teachers and his fellow students try to accommodate him by speaking up and talking to his face (he reads lips). But in a computer environment, he always needs to find a way of letting the system know, and sometimes it is very difficult. Some software packages mix music and speech in such a way that he can't understand what is being said. Animated characters don't have good enough mouth expressions to be used for lip reading. Clearly, some software developers haven't considered accommodating hearing problems in the design process.

In a computer environment, hearing and visual impairment and other perceptual needs have to be considered and taken into account from the very start. It is easy to ask if a prospective user has a preference for particular display options. For example, text size and volume should be easy to control and manipulate. But some things are just not as obvious. About twenty percent of all men are color blind to some degree—red/green confusion is the most common form of color blindness. A significant proportion are not even aware of their problem. Given such a high incidence of color blindness among the population, it seems only prudent not to use red and green cues as a way of conveying differing information. And, of course, that's just the beginning. Designing for people with perceptual disabilities is now sometimes mandated by law.

In addition to differences in perceptual acuity, people vary in their perceptual styles. Two individuals might have identical hearing profiles measured as variations in hearing plotted against audio tones, but their abilities to understand and remember audio information might be very different. And while it is possible to ask about or test hearing acuity, it is more difficult to elicit information about perceptual style preference—is it easier to learn or understand from an audio presentation or from a book? For example, some students do well in a lecture format class, but others require the teacher to write notes on the board in order to actually understand the information. The first group contains audio learners, the second does not.

Consider online learning. In this situation, where a face-to-face interaction between a student and a teacher is not a regular occurrence, it is useful to identify the perceptual styles of the prospective audience: **Auditory**, **Haptic**, and **Visual Perceptual Styles**. In a class setting, a teacher will try to present the same information in a number of different ways to answer students' questions as they arise. But most online educational materials are asynchronous —teachers and students are not there at the same time—and information is delivered in a text-heavy format with few opportunities for alternative presentations. By focusing on the different perceptual styles of the students, content developers can consciously try to make accommodations for the perceptual preferences of their audience.

The final perception category adopted by the Cognitive Wheel is **Perceptual Processing**. Here, it is defined quite narrowly and limited to the visual range of focus of attention (i.e. the size of the area in visual field that an individual is most comfortable focusing on), to the ability to process language symbols, and to social processing. These categories yield the most variation in prescriptive educational materials and product design.

Perceptual Style

A teacher who can arouse a feeling for one single good action, for one single good poem, accomplishes more than he who fills our memory with rows on rows of natural objects classified by name and form.

—Johann Wolfgang Von Goethe

Perceptual processing examines how information gets analyzed prior to being committed to memory. Perceptual processing shouldn't be mistaken for perceptual acuity—differences in eyesight and hearing. Perceptual processing occurs after information is perceived by our sense organs. In addition to audio, visual, and haptic forms of perceptual processing, there are also social, temporal, and spatial. All are defined below.

Auditory Perceptual Style

People who have an auditory perceptual style tend to understand and learn better by receiving information in an auditory format. They have difficulties following written directions. They may also have difficulties with reading and writing. They learn by listening and participating in discussions. These people prefer tapes, videos, lectures, discussions, records, radio, stereo, oral directions and explanations as a primary means of conveying information. Secondary reinforcement can be done with visual, tactile, and kinesthetic resources.

Visually presented information allows students to go over it again and again. It can also be examined in detail and be taken in as a whole—examine a tree, observe a forest. Audio, on the other hand, is time constrained—only a cross-section of information is available at any given instant. If an individual has an attention control problem, a small break in concentration can result in misunderstanding the whole audio presentation.

Auditory Perceptual Orientation: These individuals prefer to receive audio cues from their environment. They learn best when they can hear the information. Technically, Auditory Perceptual Style is part of the Haptic Perceptual Style. But for the purpose of product design specification, it's best to keep those styles distinct.

If you are an auditory oriented individual, then you:

1. Remember what you hear and your own verbal expressions
2. Remember by talking aloud and through verbal repetition
3. Desire to talk through a concept which you don't understood
4. Verbally express excitement
5. Can remember verbal instructions without recording them
6. Enjoy class discussions and talking with others
7. Are easily destructed by sound but you also find silence distracting
8. Find it difficult to work quietly for extended periods of time
9. Enjoy musical activities

Haptic Perceptual Style

The individual with haptic tendencies prefers to derive perceptual information through body sensations experienced in a tactile and/or kinesthetic mode.

Theoretically, visual and haptic learning styles are at opposite ends of a continuum of perceptual organization of the external environment. I believe that this style is highly context dependent and, as such, is not a true bipolar dimension. For example, learning to play the piano or learning to use a typewriter are both haptic activities with a strong visual and symbolic processing (language and musical notes) component.

Proprioception is the ability to sense and understand the position of one's body and its individual parts in space. An athlete, for example, clearly has a good haptic memory—the athlete's muscles remember what to do and how to do it well, with less effort than most. However, an athlete may have problems with fine motor control, as evidenced by handwriting, for example. Fine motor control and athletic ability are not linked.

Haptic Perceptual Orientation: These individuals prefer to use their bodies and hands as the primary way to learn.

If you are an kinesthetically oriented individual, then you:

1. Become physically involved in the subject
2. Enjoy acting out a situation
3. Enjoy making a product or completing a project
4. Prefer building and physically handling materials

5. Remember and understand through doing something

6. Take study notes to keep busy but often do not need them

7. Enjoy using computers

8. Physically express enthusiasm by getting active and excited

9. Find it difficult to sit still for extended periods of time

10. Enjoy hands-on activities

Visual Perceptual Style

Visually oriented individuals have a preference for using their visual cortex to process information. This is not strictly about perceiving the information through the sense of sight. Rather, it is about understanding information utilizing cognitive processes revolving around visual perception. The broad scope of visual perception includes five components: spatial relations, visual discrimination, figure-ground discrimination, visual closure, and object recognition. Not all of these skills are necessarily equally well developed in a visual individual. Visual individuals generally tend to prefer information in the form of pictures, filmstrips, films, graphs, transparencies, computers, diagrams, drawings, books, and magazines.

Visual Perceptual Orientation: These individual prefer to receive visual cues from their environment. They use their sight as the primary way to learn.

If you are an visually oriented individual, then you:

1. Desire to see words written down

2. Enjoy a picture of something being described

3. Prefer a time line to remember a historical event

4. Prefer written instructions for assignments

5. Observe all the physical elements around you

6. Carefully organize your learning or work materials

7. Enjoy decorating

8. Desire photography and illustrations with printed content

9. Remember and understand through the use of diagrams, charts, and maps

10. Appreciate presentations which use overhead cells or handouts

11. Study materials by reading over notes and organizing content in outline form

12. Enjoy visual art activities

Having a visual perceptual orientation doesn't mean that you are not capable of understanding information presented in other media. It's just that it is easier to do so visually. It's a preference, not a requirement.

Visual Perception:

- **Spatial Relations**
- **Visual Discrimination**
- **Figure-ground Discrimination**
- **Visual Closure**
- **Object Recognition**

Sequential/Temporal and Spatial Perceptual Processing

There are people that never show up on time, arriving either too early or too late; a deadline is just a suggestion to them; they have a very difficult time of making plans. And then there are individuals who live in a perpetual state of chaos—their rooms are battle zones with stuff explosively strewn all over the place; they can't read a map and can get lost on the way to a meeting; they are the ones to whom you have to say: "Your other left!" All these characteristics have to do with either sequential/temporal or spatial processing problems.

Sequential/temporal perceptual processing deals with the ability to observe and understand ordered chains of information—strings of data that have sequential patterns or a particular arrangement in time. The sequential processing system tends to be located on the left side of the brain, the spatial processing system on the right.

People with poor sequential/temporal processing have trouble in making and keeping schedules. They are always late and have poor time management skills. These individuals might also have trouble understanding a sequence of instructions, have issues with recipes, and are poor at following directions.

Spatial perceptual processing deals with the ability to figure out visual patterns and to arrange information spatially. Geometry problems, mathematical graphs, chess board positions are all examples of spatially arranged information.

People with poor spatial processing tend to have closets that look like there has been an avalanche of socks. They have trouble managing and keeping track of their possessions. Their desks are messy. And they tend to have problems reading structured visual information—graphs are difficult, tables are hard, mathematical equations seem out of reach.

There are wide differences between individuals in their abilities to interpret, store, and communicate sequential/temporal and spatial information. Clearly, if a person has poor sequential/temporal and spatial perceptual processing skills, she will have trouble remembering such information and will experience stress when forced to deal with sequential/temporal or spatial data, further burdening her already overwhelmed working memory.

The interaction with sequential/temporal and spatial information can be broken down to:

- **perceiving**—an ability to figure out a pattern of sequentially or spatially arranged data;

- **remembering**—the ability to remember a sequential or spatial pattern of information;

- **making**—the ability to create, organize, or arrange information in a sequential or spatial pattern (this includes time and materials management); and

- **thinking**—the ability to solve problems, to reason, and to think critically about sequential or spatial information.

An individual can be good at remembering sequential or spatial patterns of information, but be lousy at creating such patterns. Just as there are many more good readers than there are good writers, there are far fewer information architects than there are consumers of well organized information.

Symbolic Visual Processing

An image is not simply a trademark, a design, a slogan or an easily remembered picture. It is a studiously crafted personality profile of an individual, institution, corporation, product or service.

—Daniel J. Boorstin

Computer software has facilitated an explosion of graphically presented information. Printed material and Web pages are all teaming with graphs, tables, maps, diagrams, and charts. On top of this barrage of data for visually-inclined symbolic learners, abstract graphical content is now present in a dizzying variety of forma and media: video, animation, and other pre-visualizers of data showing information not only in a flat, 2-dimensional plane of a paper or 3-dimensions, but even sometimes with four axes. The result is a proliferation of abstract visual information, miscomprehension of which might have serious personal, professional, and civic implications. The general consensus is that "a picture is worth a thousand words." But is it really true, especially when it comes to graphically presented data?

The skill to read and comprehend visually complex information is not an innate ability. Most people get their first exposure to abstract visual information as early as kindergarten. While aspiring to more, public schools tend to teach very simple graph skills: **plot** points on a 2-D plane; **create** a bar diagram; **read** pie charts; and, in the higher grades, **graph** an equation. These tend to be **lower level visual symbolic processing skills**.

In the real world, individuals are asked to compare complex sets of data presented as charts or graphs; to interpret the meaning of visual data; and to summarize the key ideas shown in graphical ways. More importantly, in the real world, individuals have to make important personal and civic decisions based on graphical information and their understanding of its implications (e.g. personal health choices, tax decisions, election issues, etc.). These tasks require **higher order visual symbolic processing skills**. People who can't process and comprehend information presented in these formats are at a severe disadvantage.

As with any processing skill, symbolic processing can be deconstructed into receptive and expressive components: the ability to decode abstract visual information and the ability to generate it. Clearly, it's easier to read something than to write something. In particular, visual symbolic processing can be broken down into the ability to **find** information in a diagram; to **use** data in the diagram to generate new information; to **create** explanations of graphically presented information; to **compare** the usefulness of two diagrams presenting similar information; and to **summarize** the main ideas of a complex diagram.

The results of a small study I conducted in 2007-2008 revealed just how difficult it is to understand abstract graphical information. Higher level visual symbolic processing skills required in summarization, generation, and inference of new information from existing data were the most difficult and led to the largest number of communication failures. With our world increasingly relying on visual communication, this can be a big problem.

Here's an example from this study. Several of the questions were based on Inspector General Minard's famous diagram of Napoleon's March on Moscow in 1812 to 1813. Below is the first multiple choice question of the study.

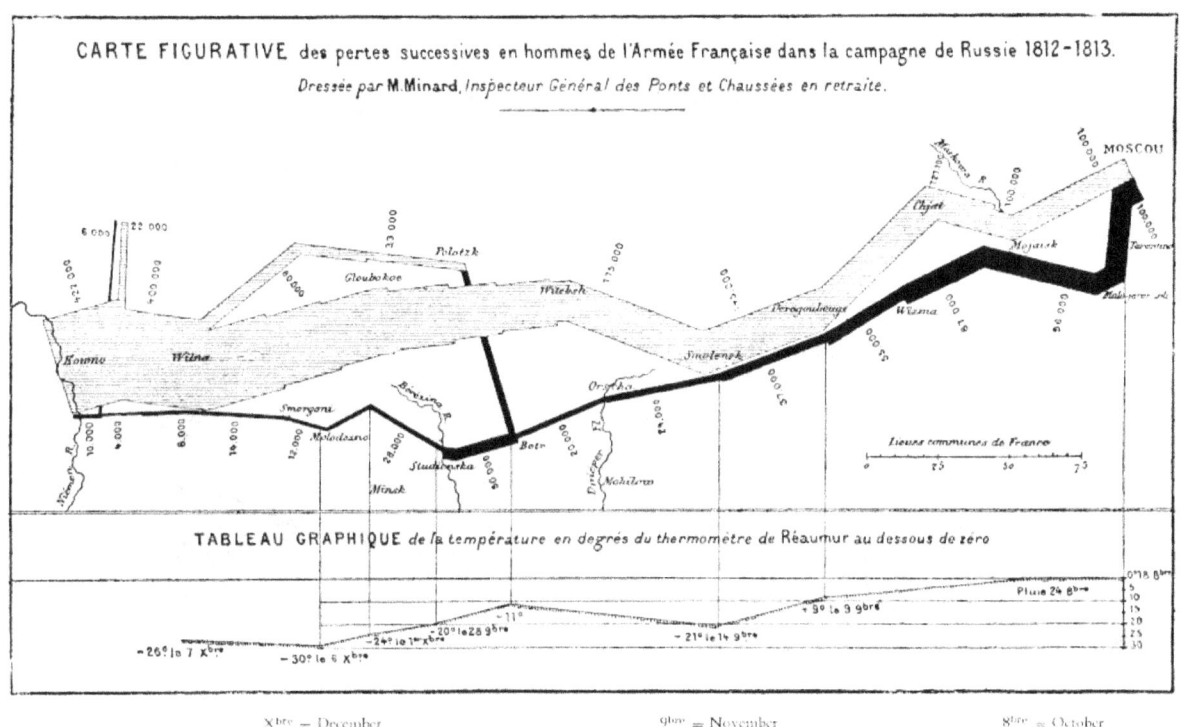

In the diagram above, Inspector General Minard illustrated the diminishing number of Napoleon troops marching to and from Moscow in several ways. Check all that you think apply.

☐ At any point in the graph above, the thickness of the line representing Napoleon's troops is in direct proportion to the number of soldiers marching in the army.

☐ At any point in the graph above, the height of the line representing Napoleon's troops is in direct proportion to the number of soldiers marching in the army.

☐ The color of the line representing Napoleon's troops indicates the number of soldiers marching in the army.

☐ The slope of the line representing Napoleon's troops indicates the number of soldiers marching in the army.

☐ The numbers next to the line representing Napoleon's troops state the number of soldiers marching in the army at that point.

☐ The line graph at the bottom of the diagram plots the number of soldiers marching in the army at that point.

☐ All of the above.

☐ This information can't be obtained from the diagram above.

☐ I don't know.

Out of 13 subjects that participated in the prelimenary study, seven answered the question correctly. Out of the five female subjects that participated, not a single person answered the question correctly.

Social Perceptual Processing

*If you have any trouble sounding condescending,
find a Unix user to show you how it's done.*

—Scott Adams

Social perceptual processing deals with the ability to recognize and understand social cues from other people. Some people are very good at this skill, and others are not.

Just like any other talent, different people are born with different abilities: some are naturally better at social skills. This is not to say that social skills can't be trained and developed, it's just easier for some than for others to achieve a particular level of social ease. And those who are fortunate enough to be born socially talented usually have difficulties understanding how others fail to excel at something that is so easy for them.

Linguistic Perceptual Processing

Language greases the wheels of perception.

—Gary Lupyan

Individuals with a **linguistic perceptual processing style** process information via linguistic symbols. This cognitive category is strictly about a person's ability to processes such data. These people are good readers and tend to have a special affinity towards written language. And while these individuals mostly use sight to perceive information, they are not necessarily good at analyzing visual information in the form of a graph or an illustration. Braille readers are also processing linguistic information, but haptically—through touch.

Consider a little test. How many F's can you find in the above illustration? Give yourself a time limit of about 20 seconds. To find out how many F's there really are, look at the end of the bibliography.

Depending on linguistic proficiency, the number of F's that a person finds in a given amount of time will vary. A young child just learning how to read, might find more F's than an adult. An adult who is just learning the language, might also find more F's than an individual with

FINISHED FILES ARE THE RE-SULT OF YEARS OF SCIENTIF-IC STUDY COMBINED WITH THE EXPERIENCE OF YEARS.

fully developed written language skills. And hearing impaired individual might find more as well. This little test shows just how fragile our linguistic perceptual processing is, and how much it depends on our background knowledge of the language.

Language and Perception

When we have difficulties finding words to describe an idea or a phenomenon, we also experience limitations in our ability to notice details or remember what we have seen. Ask a young child to describe a flower and you might get something like this: "It's big and yellow. And it smells good." Ask a botanist to do the same and you will get a very comprehensive answer. Our ability to see details is linked to our background knowledge of that subject matter.

In the article "The Forest Primeval," Peter Canby describes his journey into an African rain forest as a writer documenting a scientific expedition. While an expert wordsmith, Peter had limited knowledge of world he encountered. His realization of how this limitation impacted his powers of perception is very telling:

"A large part of my frustration comes from the language. Blake [the lead scientist] is not here to translate my questions. But I'm not just deprived of speech today; I'm also faced with the fact that the forest, which is such a source of bounty to the Bayaka [the forest Pygmies], is, to me, an undifferentiated mass. I don't have the vocabulary to break this environment into parts. There's nothing I can parse, nothing I can usefully understand. I'm completely at a loss without words. The Pygmies see that I'm wilting."

Dyslexia

It was Greek to me.

— William Shakespeare

"How many words can you process in a minute?"

Dyslexia is an example of a faulty linguistic perceptional processing. Letters and words can be brought into sharp focus with glasses, but how they get interpreted is independent of the clarity of visual information. I was diagnosed with dyslexia in fourth grade. I learned how to read a bit later than my peers, and I never learned how to spell (in any language). Till this day, I baffle the spell checker application with the creativity of my word constructions. I'm sure that in this book you will find words that are spelled correctly but have the wrong meaning. It's the result of my continued wrestling with this helpful application.

A simple accommodation would make a spell checker application a more useful tool for the dyslexic population—the thesaurus. This addition would allow for multiple pathways to information, not limited by my particular linguistic creativity.

The myth that dyslexia is just about flipping letters is just that—a myth. I do flip letters, and numbers, and musical notes, and other symbolic information like icons, but that's just a small part of it. I read differently than other people I know. I look for the overall shape of the word or a phrase and I make a guess at what it is. The longer the shape, the more likely I will get it right. We all do this with some words—the high frequency words—it's one of the ways to speed up reading and one of the reasons these words are drilled into us in elementary school. These high frequency words—it, that, this, was, is, you, he, she, go, and so on—are examples of automatic long term memory retrieval. But my reading ability strongly depends on the recognition of word shapes. It is easy for me to read as long as I stay within the subject areas that are very familiar to me—they contain many words that I can recognize at a glance. Life gets hard when I'm surrounded by novel words, especially names.

For me, driving around town looking for an address is the ultimate nightmare—not only are none of the names of the streets likely to be in my high frequency word repertoire, but the street addresses are made of strings of numbers which I know my brain will arrange in any order with little regard to reality. And I have to do it quickly as I drive past. The same is true in the airport: those strings of unfamiliar airline names and airport codes with numbers attached to them are very difficult and illusive. It's clear that airport sign designers never considered dyslexia in their usability studies.

Try comparing your reading time for these two words:

Decomposition   CRySta*l*LiZatiOn

Both words are relatively easy, but the word "crystallization" takes longer because it's hard to recognize the visual pattern. This is sort of what it's like reading unfamiliar words for me.

An example of dyslexic interference from my own experience is reading a musical score. I can read music relatively well as long as I just see the score and play the notes on the keyboard. If I have to translate the notes that I read into words, I mess up completely: not only do I name the notes I read incorrectly, but my ability to play these notes deteriorates drastically. The action of verbalization garbles the visual information that I see.

Dyslexia is not an "on and off" kind of problem, it's not like you either have it or not. It, like most cognitive characteristics, is a continuum—some people have it worse than others. Mine is not very bad. Dyslexia also doesn't stay stable across situations. Again, like other cognitive characteristics, it's situation dependent. Under high stress (tests and airports), my dyslexia gets worse.

There are estimates that as much as 10% of the U.S. population might have dyslexia to some degree. That's a huge user group undeserved by the product design community. Any time you see text that is in all caps, consider that the shape of that text hides rather than reveals information to a dyslexic individual. And when designers get cute with their naming conventions, they are unwittingly create linguistic traps for those of us for when linguistic perception processing is a challenge.

Audio Perceptual Processing

On first take, one might guess that, given equal hearing potential, two individuals would hear the same sound. But in fact the capacity for hearing is only one part of the equation. We briefly discussed perceptual styles—some individuals prefer to get their information through visual means, others through language, and some want it in an audio format. But once the information is in, how it gets processed depends on the background knowledge of the person. Consider the sounds that particular animals make. Those sounds don't generally change depending on the country that the animal is located. But how those sounds are heard and interpreted does depend on the cultural background of the listener. In Russia, dogs make "gav-gav" sounds. In Japan, they go "wan-wan." In Korea they express themselves with "mung-mung." And in Italy dogs say "bu-bu." How do you hear the sounds that dogs make?

Audio Processing Disorder

Some sounds come at us at a leisurely pace: o, u, e. These sounds are easy to hear and process. There are others, fast consonants, that hit us fast and hard. These sounds are more difficult to process. The problem is not hearing—a person might have a perfectly acute hearing and still not be able to distinguish between a "p" sound and a "b" sound. The problem is speed of processing—sometimes, the brain's audio processing speed is just too slow to pick out fast-paced phonemes, resulting in phonological awareness difficulties. So instead of hearing "dad" a person might hear "bad." The first sound gets lost, and individual ends up guessing what it was. This clearly results in audio comprehension issues, spelling problems, and even hampers thinking.

The computer is particularly good at helping people with audio processing disorders. These fast sounds can be artificially slowed down for easy comprehension. With drill and practice, the brain can be trained to recognize and process the fast phonemes, resolving the problem completely. Unfortunately, a lot of people are not even aware they have this problem, instead thinking themselves to be stupid or slow when the actual problem lies in audio processing.

Audio Processing and Cultural Differences

There are multiple "k" sounds in the Eskimo language. While any six-month-old infant can hear the difference, adults and older children from non-Eskimo cultures are at lost to identify a particular one. At birth, all humans have the ability to hear and differentiate the sounds of all human languages. But after the first few months of initial exposure, infants specialize to process only the sounds of the language spoken by their family members. After almost two decades of marriage, my husband still can't pronounce my name as it was given to me at birth. The soft "l" sound of the Russian language is just too difficult for an English speaker.

The individuals who learn additional languages later on in life, have more difficulty processing audio information of those new languages. Most non-native speakers retain accents, however slight, due to this inability to correctly process the sounds. And just as the audio processing disorder, **oral non-native language processing difficulties** can lead to miscomprehension of spoken information and spelling problems.

In our day and age, product designers have to take into account communication problems that arise from oral non-native language processing difficulties among their audience of users. The set of sounds that cause problems in comprehension between any two major languages is fairly well know and is easily researched.

Perceptual Blindness

You can only predict things after they've happened.

—Eugene Ionesco

Daniel J. Simons of the University of Illinois and Christopher F. Charbris of Harvard University conducted an experiment designed to test our ability to process visual information (as opposed to our ability to see using our eyes). They asked a group of students to watch a video of a group of basketball players passing the ball to each other. The students were instructed to count how many passes were made during a certain period of time. After 35 seconds, a man in a gorilla suit ran into the field of players, beat his chest, and ran out of the room. When the researchers asked the students whether they saw a gorilla, 50% said no!

How could a group of visually healthy students miss the appearance of a man in a gorilla suit? They didn't expect a gorilla, and so it slipped past their perceptual processing, which was occupied with counting the basketball passes. In fact, when these students saw the video again, they accused the researchers of switching tapes!

It's good to keep in mind just how much of an impact our expectations have on our ability to control our attention controls and to process information. This is particularly relevant in analyzing eyewitness testimony in criminal cases.

To read more about this experiment, visit http://viscog.beckman.uiuc.edu/djs_lab/

Visual Perceptual Processing

What do you see in the checkered image?

Do you see the circle floating above the plane? Do the black and white rectangles inside the circle move? This image is a simple optical illusion: a one cell animation. Does it make you dizzy? While these shapes and colors are simple, our brain processes this visual information to include movement and depth.

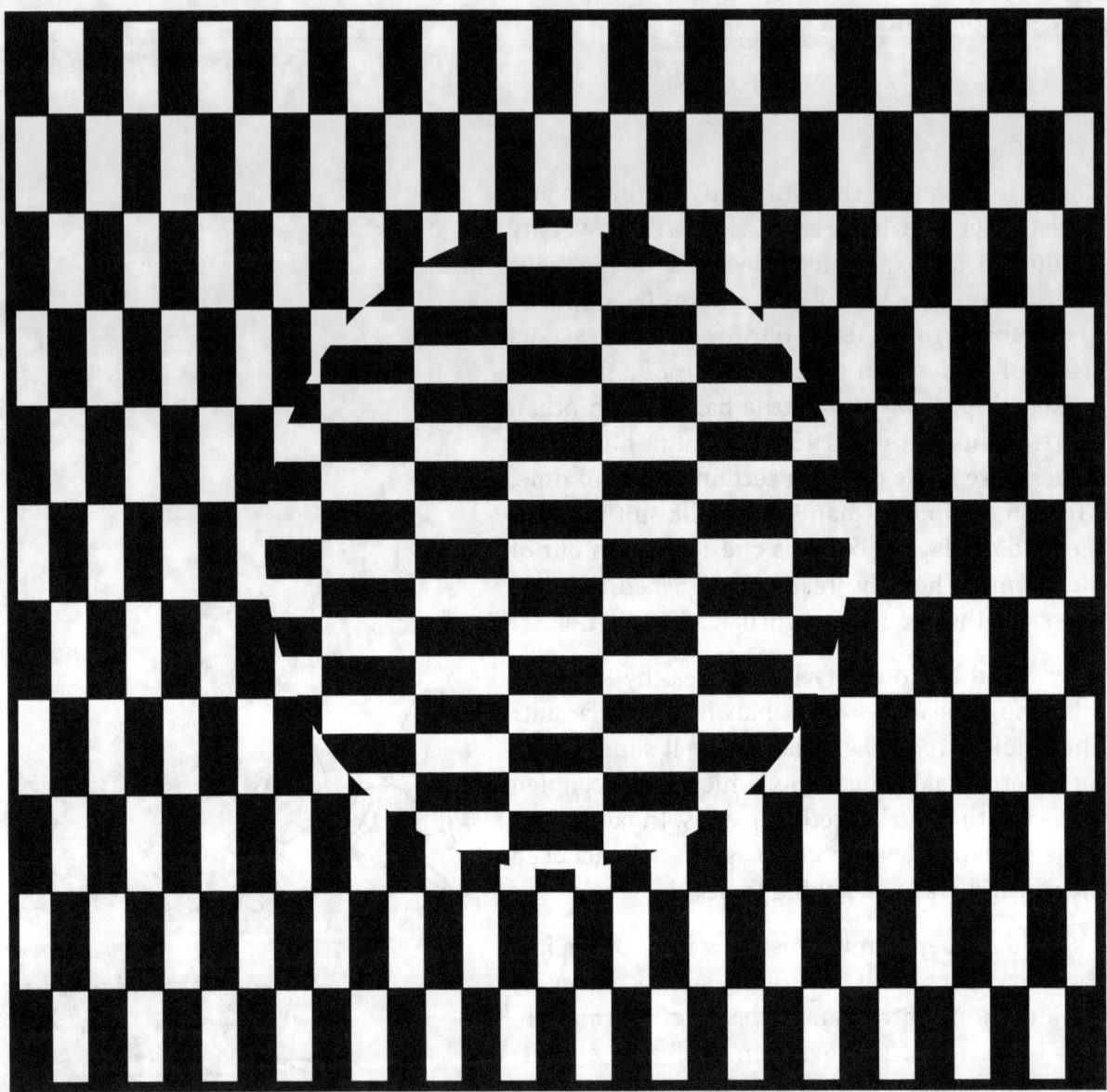

Now take a look at one of the Rorschach Inkblots used by psychologists to check individuals' visual perceptual processing responses against the norm. Over many years, the data collected was used to extract supposed correlations between violent tendencies and what people think they see in these images.

What do you see in this ink spot? According to Hermann Rorschach (1884-1922) if it's something other than a bat or a butterfly, you'd better not say!

Perceptual Processing Level of Detail (Focuser/Scanner)

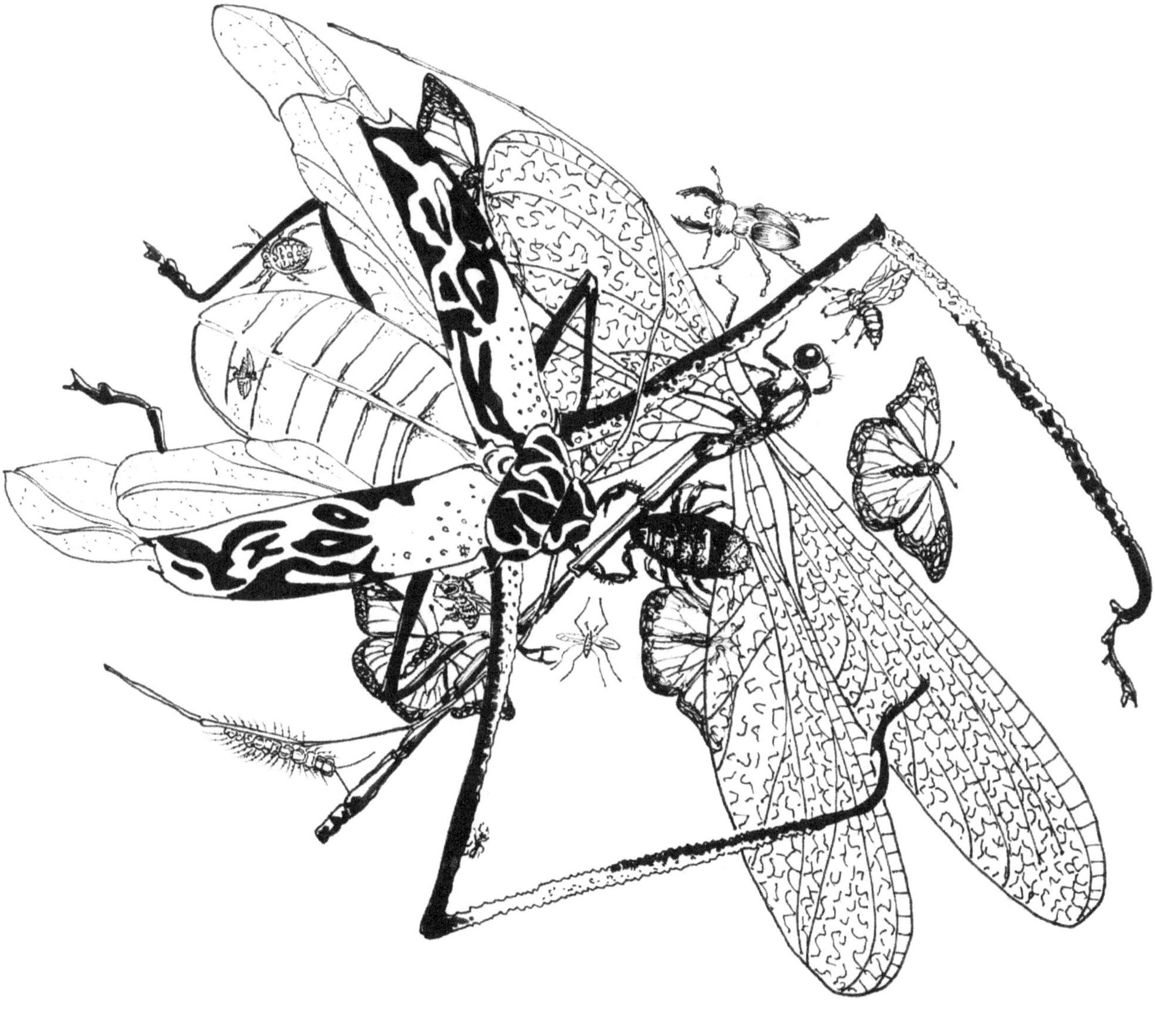

This cognitive category describes individual differences in the processing of a stimulus field. Generally, when individuals scan a field, they cognitively record and compare both visual and verbal properties from the available information. **Focusers** tend to get bogged down in detail, while **Scanners** are able to do higher-level cognitive processing of material.

By definition, these are two ends of a continuum. At one end are the Scanners. These individuals direct their attention actively and freely to all parts of the field. They miss a lot of detail and have a wide range focus of attention. At the other end are the Focusers. These individuals are passive and direct their attention to a very narrow field of information. They have a narrow focus of attention and a restricted attention to fewer facets of their surroundings, but they notice more details from that limited field.

Count how many insects you see in the pen and ink drawing. The answer is in the Bibliography.

Additional Thoughts and Further Readings

I wrote an article for Ed-Media 2008 that presented the results of a small study examining the issue of abstract graphical understanding. The paper can be download from my company's Web site: www.pipsqueak.com.

Dr. Sacks' book "The Island of Colorblind" is sheer poetry, you don't have to be interested in visual perception to find it wonderful.

If you have a bit of curiosity about this subject, but not enough to read another book, consider the illustrations in this book. It is cheaper to print, and thus to buy, a book without color illustrations on the inside. Black and white images (grayscale) clearly limit the information that we perceive, or do they? Color can make some information jump out of the page. That same color can be invisible among the tones of the overall illustration. But color can also be used to camouflage information (thus the colorblind people on the island of Pohnpei are better, counter intuitively, at identifying ripe fruit). As an experiment, compare color versions of illustrations on the cover of this book with those on the inside. Is there different information that jumps out at you?

Dr. Sacks wrote more than one book about perception and cognition. I would strongly recommend the following books: "A Man Who Mistook His Wife for a Hat," a collection of essays on unusual manifestations of neurological conditions; "Seeing Voices," a history of treatment and mistreatment of individuals born with no or limited hearing (this is also a wonderful book for those interested in culture differences between groups with varying perceptual abilities); and "A Leg to Stand On," a book documenting proprioceptional difficulties that Dr. Sacks experienced after an unfortunate personal encounter with an angry bull. The last book was particularly revealing to me—it is the first account I read of someone (other

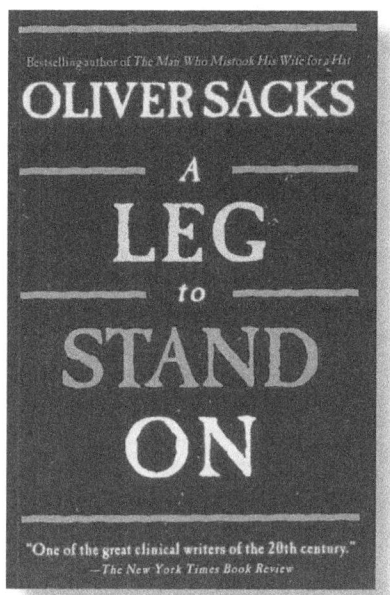

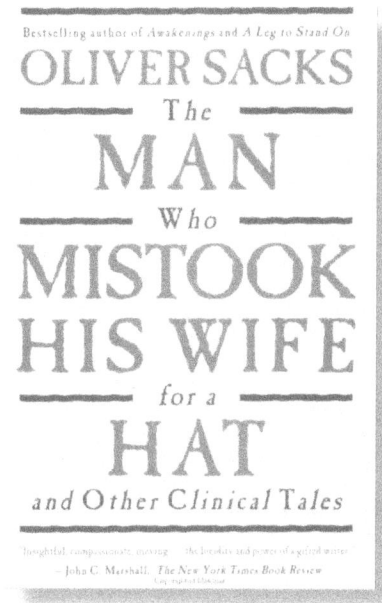

than myself) describing their leg as having a "will of its own." Many years ago, I was run over by a taxi while crossing the street as a pedestrian. I daily use a cane and I still encounter that "not exactly my own" feelings about my left leg.

For more information about audio processing, visit www.scilearn.com.

If you're interested in checking out a few more visual puzzles, visit www.exploratoruim.edu— the Web site of the coolest science museum in the San Francisco Bay Area.

13. Language and Linguistic Processing

Handle them carefully, for words have more power than atom bombs.

—Pearl Strachan Hurd

Language is one of the first tools invented by humans. The ability to pass information from one individual to the next, from one social group to the next, and from one generation to the next is essential to our survival as a species. But language is not only an inter-personal tool, it's one of the main tools we use during problem solving—we "talk ourselves through" a problem to get to a solution; we formulate goals and keep them in mind using language; we form plans and devise strategies; we generate explanations using language; and we mediate our emotions with language. And a lot of our memories hang from a backbone constructed of language as opposed to one constructed of images, for instance. We depend a lot upon language, although not exclusively—there are visual, kinetic, and audio tasks that have a "language" of their own and can be pre-visualized without the use of words. Think of dance movements, music composition, painting and sculpting—all of these activities can be accomplished without words.

To talk about language, we need a vocabulary to express linguistic concepts and to differentiate between different types of language that we encounter daily.

Language Components

At the bottom of the language pyramid are phonemes. **Phonemes** are sounds which form the basis of a particular language. When we are born, we are capable of distinguishing sounds from all human languages. But by about six months of age, that ability to distinguish individual sounds diminishes, leaving each of us permanently predisposed to hearing the sounds of the particular language that we have been exposed to since birth. English has 44 phonemes.

Morphemes are the building blocks of words. They are strings of sounds from which the words are constructed. The sound that "tion" makes in the words "vacation, transportation, relaxation" is an example. The English language contains morphemes from many other languages with which English speaking populations historically came into contact, particularly Latin, Greek, and Old French.

Semantics is the study of the meanings of language and its words. By studying the history of words and their evolution through time, we have a better grasp of the meanings of those words. By understanding the composition of words that have been made of other words, we can recognize and understand the meanings of the words that we haven't previously encountered. Individuals with large personal vocabularies not only know more words than an average person, but they understand the meanings and use of those words better than the average person—they are skilled users of words.

Syntax is the rules of a particular language for stringing words together to form sentences and the effect that word order has on the overall meaning of a sentence.

Discourse is language in big chunks.

And finally, **metalinguistics** is the study of language generally.

The ease with which an individual interacts linguistically within a culture depends on:

- the ability of the brain to detect and differentiate the phonemes of the language;

- the number and degree of understanding of the different morphemes—a person with a poor grasp of morphemes would have significant problems with spelling, for example;

- the individual's vocabulary—the number of words the person knows how to use, the quality of the person's knowledge of the definitions of those words, and how well the words spring to mind when required;

- the knowledge of syntax—the ability to express oneself using either verbal or written language; and

- the speed of language comprehension—the ability to process written or spoken information at the pace that it is being delivered.

Taxonomy of Language Usage

In order to further understand possible language difficulties, it helps to have a taxonomy of language. The most obvious division is between receptive and expressive languages. **Receptive** language is the understanding of spoken and written language directed at an individual. **Expressive** language is the individual's ability to communicate linguistically, either verbally or in written form. People whose first language is not English often complain that they understand more than they can express—their expressive language abilities lag behind their receptive ones. Fortunately for those individuals, they have another language to fall back upon. A person who speaks only one language and has expressive language problems would experience difficulties talking to herself while solving a problem, for example. For these people, the internal dialogue that we all carry with ourselves is just not robust enough to be of much help in generating either a solution or an explanation.

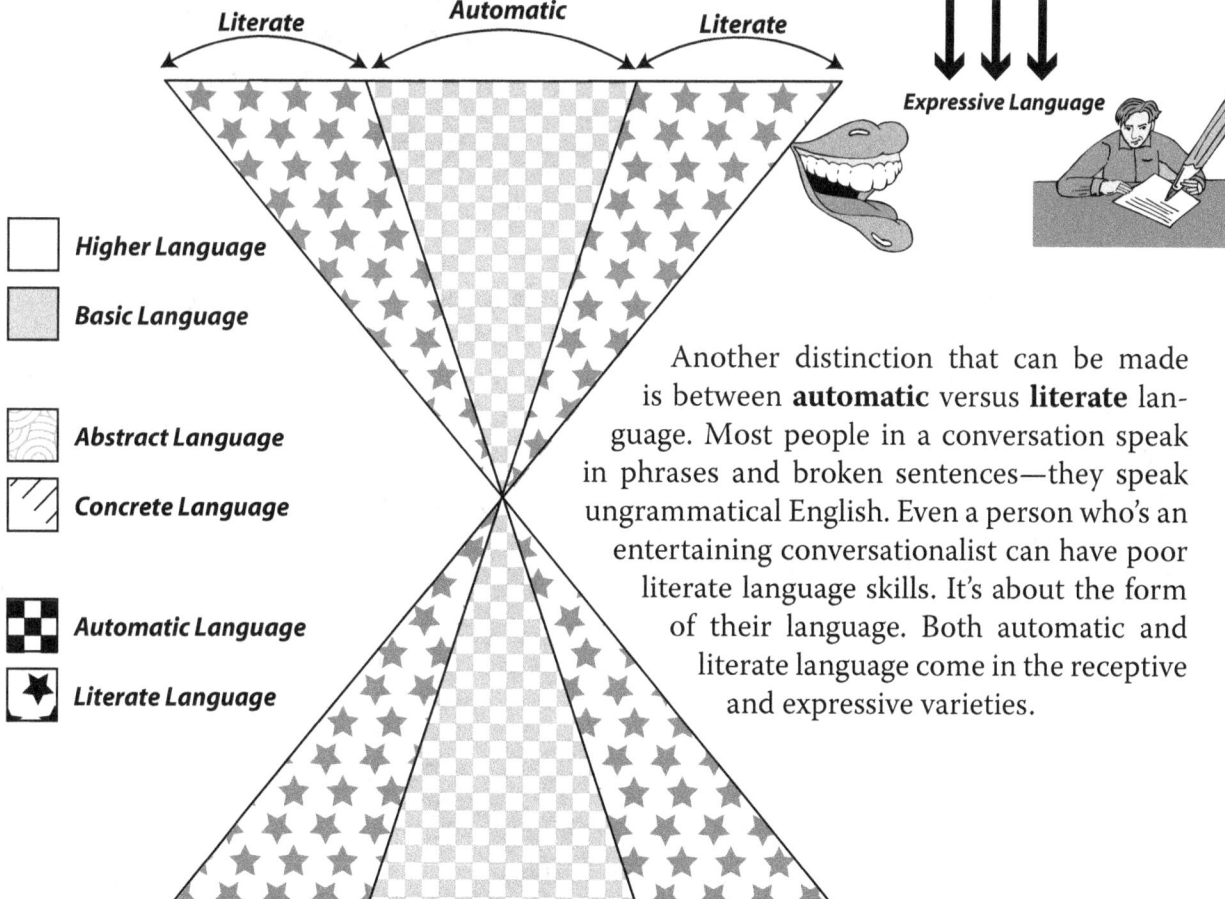

Receptive Language

Expressive Language

Literate   Automatic   Literate

☐ **Higher Language**

▨ **Basic Language**

Abstract Language

Concrete Language

Automatic Language

Literate Language

Another distinction that can be made is between **automatic** versus **literate** language. Most people in a conversation speak in phrases and broken sentences—they speak ungrammatical English. Even a person who's an entertaining conversationalist can have poor literate language skills. It's about the form of their language. Both automatic and literate language come in the receptive and expressive varieties.

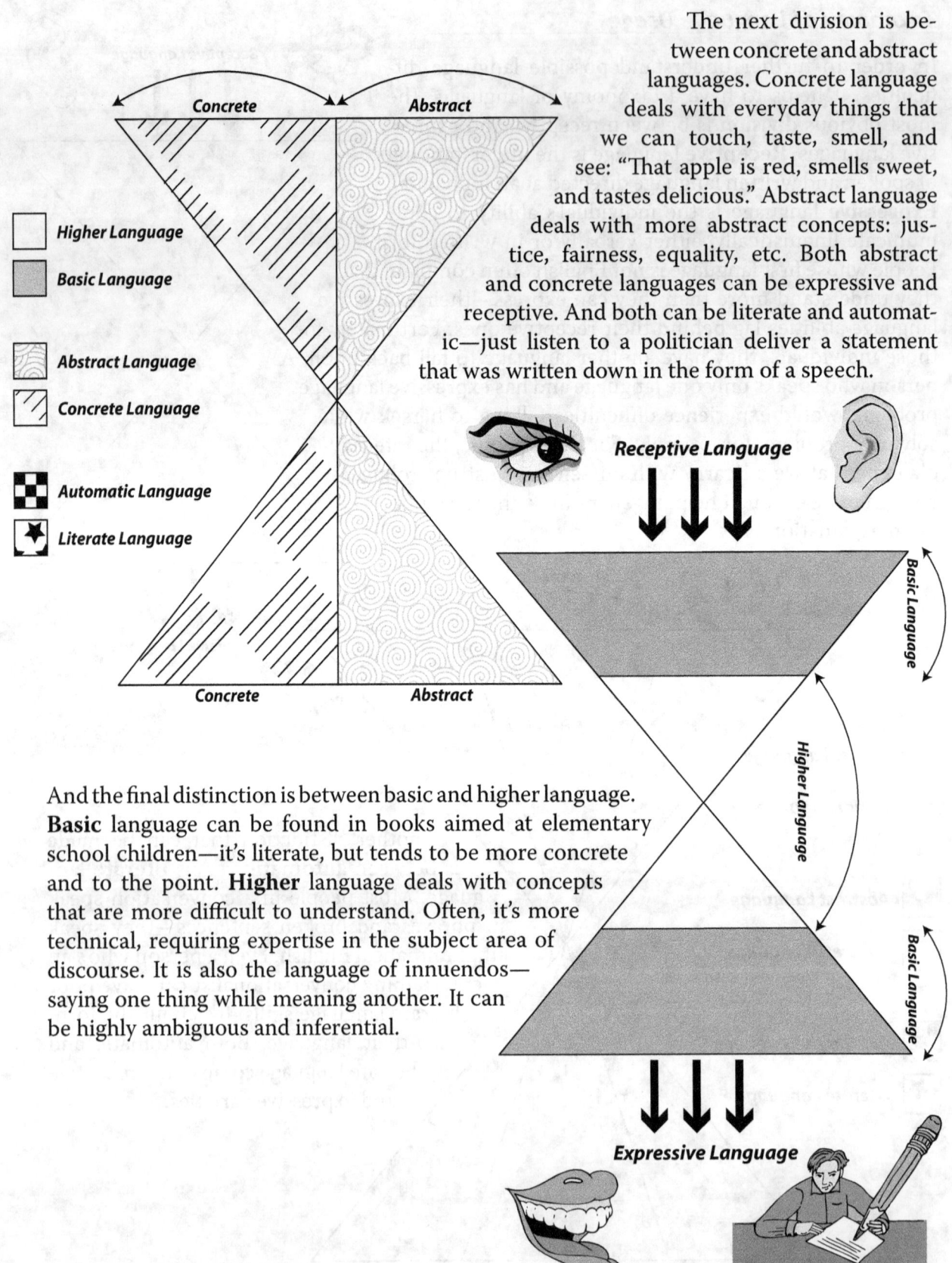

The next division is between concrete and abstract languages. Concrete language deals with everyday things that we can touch, taste, smell, and see: "That apple is red, smells sweet, and tastes delicious." Abstract language deals with more abstract concepts: justice, fairness, equality, etc. Both abstract and concrete languages can be expressive and receptive. And both can be literate and automatic—just listen to a politician deliver a statement that was written down in the form of a speech.

And the final distinction is between basic and higher language. **Basic** language can be found in books aimed at elementary school children—it's literate, but tends to be more concrete and to the point. **Higher** language deals with concepts that are more difficult to understand. Often, it's more technical, requiring expertise in the subject area of discourse. It is also the language of innuendos— saying one thing while meaning another. It can be highly ambiguous and inferential.

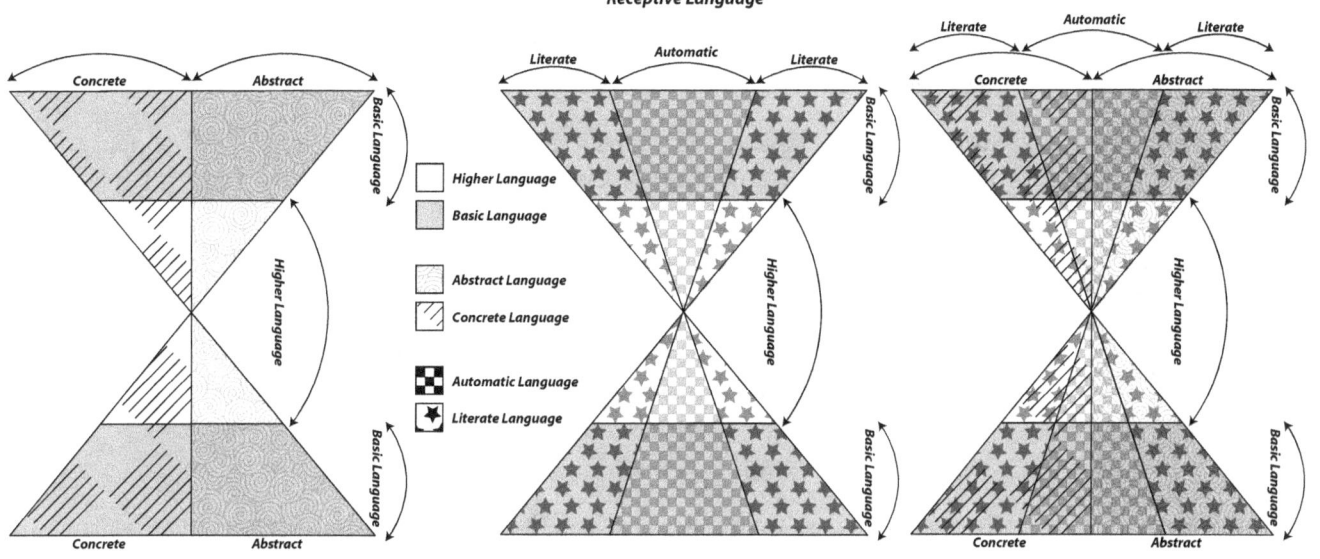

Degrees of Knowing English

If Gore invented the Internet, I invented spell-check.

—J. Danforth Quayle

English is the official language of forty-four countries. It is estimated that there are about 330 million native English speakers in the world, and 400 million use English as their second language. Globally, there are 200 million people who speak English-based creoles. These people use English, but do they know English? What does it mean to know English?

After 25 years of living in America, I have a certain degree of competency in English. But not so my parents—they get by, but only barely. Their receptive and expressive language skills in English are still very poor. They can get only basic ideas across and even those sometimes suffer in translation especially when they try to tell jokes. My parents are not stupid people, it's just that they lack competency in the English language. Yet, if you ask them (or others who know them), they would say that they can speak English. But clearly my English and their English are not the same—we have different degrees of knowing this language. And in America, there are over 40 million people living here who do not speak English at all!

It is not only that non-native speakers of English introduce a variety of levels into what it means to know the language. Consider the West Coast and the East Coast of the United States. People in New York tend to speak faster than people in San Francisco. But it's not just the speed of the language delivery that is different. It's the difference in vocabulary and the turn of phrase (not to mention the accent) that jump out. And how about England and United States? Do we speak the same language? How about the English spoken in Ireland or Scotland? After years of living with us, our Irish baby sitter still managed to surprise us with

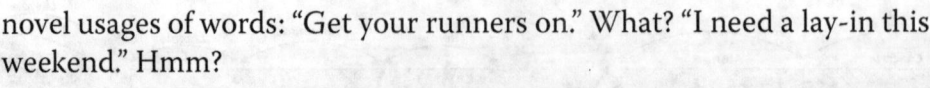

novel usages of words: "Get your runners on." What? "I need a lay-in this weekend." Hmm?

When designing a product for an English speaking audience, consider carefully that audience's ability to speak the language. And on this note, here is a quote from our former Vice President, Dan Quayle:

> *"I suppose three important things certainly come to my mind that we want to say thank you. The first would be our family. Your family, my family—which is composed of an immediate family of a wife and three children, a larger family with grandparents and aunts and uncles. We all have our family, which-ever that may be."*

Mr. Quayle was trying express his thoughts at a Thanksgiving festival in Charles City, Virginia, in 1988.

Reading Strategies: Deep vs. Surface Reading Strategy

It's not enough to be able to parse through a few sentences or even a book. Reading is about comprehension, and this ability varies. **Deep reading processors** think about the author's purpose and relate each reading assignment to prior knowledge. **Surface processors** read with little thought. Surface processors do poorly in reading comprehension. Deep reading is easier in one's areas of expertise, where an expert can see broad patterns in information as opposed to focusing on understanding individual details.

Unfortunately, we are currently a culture of surface readers. And surface editors:

What service do you think is being offered by a "passionate" caregiver charging by the hour?

The Power of Language

I've learned any fool can write a bad ad, but it takes a real genius to keep his hands off a good one.

—Daniel J. Boorstin

There is enormous power in language. It lies not only in the ability to precisely express an idea, but also in its ability to change another person's mind. An obvious example of such language use is an argument—a logical sequence of a string of ideas which have the power to shift another's position. An argument is an expressive language art.

Now consider a not so obvious use of language: "Tax Relief." What does this two word phrase mean to you? It's a very powerful phrase meant to manipulate the minds and hearts of American voters. By adding the word "relief" to the word "tax," the meaning of the word "tax" changes. To relieve someone of something is a positive thing. The word "relief" is commonly coupled with medical conditions: heartburn relief, headache relief, pain relief. In our culture we also have ideas like: "debt relief" and "mortgage relief." Clearly relief from those financial responsibilities is a welcome thing. By attaching the word "tax" to "relief," we can create a carefully designed emotional response from an audience: tax is a bad thing and you want to be "relieved" from it. But the manipulation doesn't stop there. Not only is "tax" bad, but the person who provides you with "relief" from it is a good guy, doing something positive for you. Next time you hear this phrase, think how you are being manipulated and by whom.

Language and Society

There are approximately 6,000 languages spoken in the world today. When trade, cultural exchange, education, and diplomacy would be greatly facilitated with a single language, why do humans use so many? At the dawn of humanity, some 50,000 years ago, when our population numbered in just a few thousands individuals, we probably spoke just one language—the mother tongue. The study of language evolution, as it paralleled human evolution and migration throughout the globe, is a fascinating story of this very basic and fundamental human tool.

Language fragmentation serves as a barrier—it's the easiest way to identify members of a particular group. Throughout history, people had a strong incentive to invent and use words that a hostile (or just different) community members would not understand. Language differences served to keep communities safe during continuous warfare between neighboring states. If an individual couldn't pronounce a word correctly or used a wrong expression, then he might have been an unwanted intruder—a spy. The Sicilian Vespers—the massacre of 1282—used a linguistic challenge to identify and murder the occupying French troops amidst the Sicilian population. The word was "chickpeas."

Language variations serve to identify members of different **geographic** areas and **political** groups. And they are used by members of one generation to differentiate themselves from previous ones, creating **intergenerational** linguistic variations.

Additional Thoughts and Further Readings

I write to discover what I think. After all, the bars aren't open that early.

—Daniel J. Boorstin

In his book "The Mother Tongue," Bill Bryson provides wonderful descriptions of English as it is used and abused (or should we say "made their own") by people around the world. Consider an advertising message for a company called Cream Soda: "Too fast to live, too young to happy." Did their advertising executives speak English? How about a can of Coke from Japan: "I feel coke & sound special."

For those interested in the origins of phrases like "say uncle," check out Daniel Cassidy's "How the Irish Invented Slang." This is a fascinating read for English speakers.

And if you find politics entertaining, you might want to check out how language is carefully chosen to make you believe and trust a particular politician or a campaign slogan. "Safire's Political Dictionary" by William Safire will change the way you listen to political speeches and the news, for that matter.

Another book that highlights the use of language in politics is Geoffrey Nunberg's "Going Nucular: Language, Politics, and Culture in Confrontational Times." You will be thoroughly manipulated by the end.

If you love words, check out www.americandialect.org. The section on the "words of the year" gives an interesting look at cultural pressures on language evolution. "Googlegänger," "NINJA," and "wide stance" are direct reflections of 2007 American society. Just as "9-11," "weaponize," "ground zero," and "women of cover" were words and expressions that linguistically reflected the mood of 2001. There's a lot of emotional charge behind each of these words, and they bring those emotions to conversations which use them.

Denis Mack Smith's "A History of Sicily, 800-1713: Medieval Sicily" has a complete account of the Sicilian Vespers.

14. Attention & Impulsivity Controls

The greatest pleasure in life, is doing the things
people say we cannot do.

— Walter Bagehot

"Go get ready for bed: grab your snacks for tomorrow's lunches and put them on the kitchen table, put your dirty clothes in the hamper and put on your pajamas, brush your teeth, take an allergy pill." We've been saying this mantra to our children for the last five years. Our success has been variable. Once, when my mother was visiting us, she inquired what they could possibly do up there for forty-five minutes? As I called them down, one of my sons was only partially dressed, neither had brushed his teeth or taken the pill. They said they had forgotten—is this an example of memory difficulties? I think not. Rather it is an example of attention control failure. They were too busy doing something else and my bidding slipped their mind. There are other circumstances when the distinction between attention controls and memory difficulties is not as clear. Below are descriptions of attention controls and how their failure might manifest itself.

The Flashlight Metaphor

If working memory is like a writing desk where all cognitive work is done, then attention controls can be thought of as a flashlight that moves a spot of light around that desk. If you're tired, the flashlight is dim and erratic. If you have an attention disorder problem, then your flashlight jumps around from object to object on your desk and the size of the cone of light hitting the desk changes in size, diminishing and increasing without much direct control from you.

Attention Controls:

- **Mental Energy**
- **Intake**
- **Output**

There are three main types of attention controls: **mental energy controls**, **intake controls**, and **output controls**.

Mental energy controls describe the differences between individuals' abilities to focus intensively on a task for a period of time. Some people need a lot of little breaks while they study or perform cognitively difficult tasks, and some don't. Some individuals sleep well at night and some don't. Everyone is different.

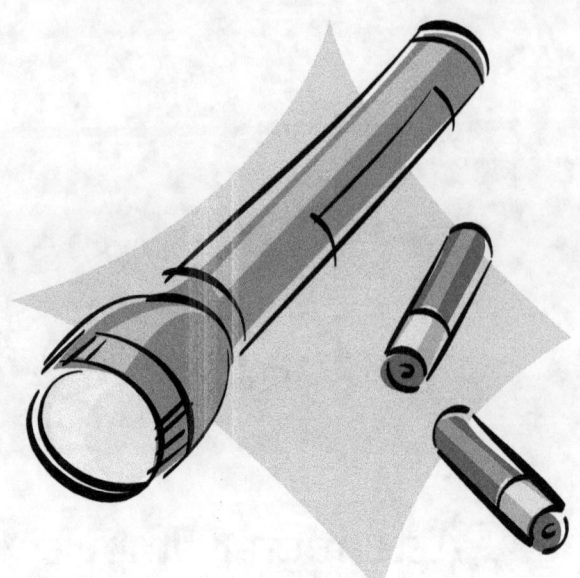

Intake controls describe individual differences in acquiring new information. Some people are seemingly born knowing how to work and study. Others have to learn how. Some people understand what's important and can focus on just the right ideas for just the right amount of time. Others have more difficulty finding informational cues. Some get enormous gratification out of getting every problem right. Others don't derive much pleasure from the effort.

Output controls describe the differences in individual work product. The same mental effort can result in a prolific output or a scant dribbling of work. Some people talk about working but produce very little. Just spending a lot of time is not enough.

A lot of these differences can be explained with varying attention controls. But it's important to note that a particularly limited short term memory, for example, can look very similar to a breakdown in intake controls. And extensive expertise and interest in a particular subject can compensate for scarce mental energy as well as difficulties with intake controls.

Mental Energy Controls

Continuing the flashlight analogy, **Mental Energy Controls** can be thought of as the amount of light that shines from the flashlight. Mental Energy Controls are made up of **Alertness Control**, **Mental Effort Control**, **Consistency Control**, and **Sleep-Arousal Control**. Some children with mental fatigue become hyper, using excess physical energy to compensate for their lack of mental energy.

The amount of mental energy an individual has at any given time varies during the day. Some people have more in the morning, but others perk up after lunch or even after sun's down. As adults, we have an intuition for what is the best time of day for us to do difficult tasks that require a lot of concentrated mental effort. Children are locked into a school schedule and have very little influence in correlating their "best time of day" to the cognitive demands of school.

Mental Energy Controls:

- **Alertness Control**
- **Mental Effort Control**
- **Consistency Control**
- **Sleep-Arousal Control**

Sleep-Arousal Control

You can think about **Sleep-Arousal Control** as the rechargeable batteries in your flashlight. If you don't recharge the batteries, you don't get much light; if you don't get enough sleep, you don't have much energy to think. A child with sleep-arousal control problems has difficulties falling asleep, staying asleep, waking up in the morning, and getting enough sleep.

Consistency Control

Consistency Control is like the wiring system in the flashlight—your batteries might be just fine, but the wires are faulty and the light flashes on and off or dims and brightens erratically. If you have consistency control issues, you have enough mental energy but can't control it. A person with consistency control problems has difficulties reliably focusing on a task, sometimes generating great work and sometimes unable to focus on getting the job done.

Mental Effort Control

Mental Effort Control is the quality of the batteries—if the batteries can't hold their charge for long, the light can shine for only small periods of time. If you have mental effort control issues, you have mental energy but it doesn't last long. A person with mental effort control problems has difficulties staying focused on her work for long periods of time, although cases of extreme interest—like video games, for example—seem to make the pattern disappear.

Alertness Control

Alertness Control allows you to monitor your flashlight. If you have alertness control issues, you have problems noticing if you are paying attention. A person with alertness control problems has difficulties knowing when he is not paying attention—daydreaming, fidgeting with objects or hands, doodling, etc.

Intake Controls

Intake Controls:

• Selection Control

• Depth and Detail Control

• Mind Activity Control

• Satisfaction Control

• Span Control

Intake Controls are made up of **Selection Control, Depth and Detail Control, Mind Activity Control, Satisfaction Control,** and **Span Control.**

Selection Control

Selection Control deals with choice—what do you need to point the flashlight at? A person with poor selection controls has difficulties recognizing what to pay attention to, what to focus on, and what is important. This is particularly problematic in subject areas where an individual doesn't understand the material too well. An expert can easily focus on the heart of

the matter, while a novice flounders and wastes energy on extraneous content. This is one of the defining characteristics distinguishing an expert from a novice—experts know what to point their flashlights at.

Span Control

Problems with **Span Control** result in an inability to focus on the right task for the right amount of time—the flashlight zooms about the table and doesn't illuminate the right object for the necessary amount of time. A child with poor span control has problems judging difficulty levels of a particular task and marshaling resources for the necessary effort.

Satisfaction Control

Satisfaction Control deals with personal gratification. For some individuals, the flashlight only shines on the most exciting objects. There are two kinds of insatiability: the need for extreme experiences—think extreme sports and roller coasters—and the need for acquiring more and more material possessions. A person who has satisfaction control problems always wants and demands more.

Mind Activity Control

Mind Activity Control deals with the ability to relate current experience with previously learned information. If stacks of books in a library represent long term memory, then mind activity control informs the librarian about new books and assists with the creation of a good card catalog. A person with mind activity control problems has difficulty stepping back and monitoring progress through an activity and relating that activity to past experiences.

Mind activity control, like other good learning strategies, can be explicitly taught, although it rarely is in school. Sometimes, a diary might serve as a vehicle for learning mind activity. Or having worksheets that ask questions like: "Where have you seen this pattern before?" "Do you recognize a similar concept/idea in another subject area?" "How does/could this relate to your daily life?"

Depth and Detail Control: Focusers vs. Scanners

Depth and Detail Control can be thought of as the radius of the light cone—how much of your thinking desk is illuminated at one time? A person with poor depth controls may have problems retaining information in short term memory, focusing on too many details and missing the grand organization. Alternately, an individual might just see the forest, so to speak, and miss all the trees, such a person would have problems on multiple choice or fill-in tests, for example, but might do well on essay questions.

Some flashlights have a very wide but dim cones. Some use the same amount of energy but focus it narrowly and brightly.

By definition, there are two ends of a continuum. At one end are the **Scanners**. These people direct their attention freely to all parts of the perceptual field. But they miss a lot of detail. They have a wide focus of attention—a wide but dim cone of light.

At the other end are the **Focusers**. These individuals direct their attention to a very narrow field of information. They have a narrow focus of attention and confine their attention to fewer facets of their surroundings. These people notice more details from this limited field. This is a bright pin point of light.

It's good to be somewhere in between—not too wide, not too narrow. You want an attention system capable of perceiving the big picture without losing too much detail in the process.

Output Controls

Output Controls:

- **Options Control**
- **Pace Control**
- **Previewing Control**
- **Quality Control**
- **Reinforcement Control**

Output Controls are comprised of **Options Control**, **Pace Control**, **Previewing Control**, **Quality Control**, and **Reinforcement Control**. Individuals with output controls failures tend to produce low quality work, both on the job or at school, and at home. At school, teachers tend to write the following comments on papers produced by students with output failures: "Sloppy," "Needs Work," "Edit," "Missing Conclusion," "Late," and "What are you trying to say?"

Reinforcement Control

Reinforcement Control is related to long term memory—did this work in the past? It deals with the ability to remember and to relate information. People who have issues with reinforcement controls have problems learning from their mistakes as well as their successes. Both negative and positive memories of events can serve as reinforcement triggers. Winning an award for writing the best news article creates positive reinforcement, encouraging the reporter to do it again. Similarly, getting a low evaluation at work or school tends to serve as a negative reinforcement—an individual with strong reinforcement controls looks for ways to avoid the same situation by either working harder or avoiding the assignment. Reinforcement control is related to satisfaction control.

Quality Control

Quality Control deals with self-regulation and self-monitoring. A person that has quality control problems has difficulty recognizing the poor quality of his work. He might also not realize that he doesn't understand something due to a failure to monitor his own comprehension.

Previewing Control

Previewing Control deals with the ability to recognize consequences, to look ahead, plan, and make predictions about outcomes. People with previewing control issues have problems estimating answers, for example, and thus have difficulties checking their mathematical results to decide if their answer is somewhere in the ball park. An individual with previewing control problems will also find it difficult to gather all the materials necessary for doing the job and will thus spend a significant amount of time looking for stuff and interrupting their work flow.

Pace Control

Pace Control deals with the ability to regulate the pace of the activity. A person with pace control rushes through work and makes lots of careless mistakes. Such a person has a difficult time going back and editing or correcting her work.

Options Control

Options Control is the ability to recognize multiple ways to proceed. A person that has option control difficulties acts impulsively and does the first thing that comes to mind during problem solving, ending up doing things the hard way. Such people have difficulties previewing and weighing options.

Option Control is very hard for kids. It's difficult even for adults. Experts in a particular subject area show their expertise by wielding fine options control—they know multiple way to tackle a problem and quickly realize when they start down the wrong path. Options control requires knowledge and information. Thus for novices, options control is more difficult. But options control can be developed through both expertise and study and work habits.

It should be easy to see how important attention controls are to product design. Most errors made during interaction with a product are due to poor attention controls. It is critical to understand and develop good support structures that will prevent or minimize failures due to attention controls.

Additional Thoughts and Further Readings

The desk and the flashlight metaphor for attention controls works well in the context of product design. For a different metaphor using plumbing, read Dr. Mel Levine's excellent book "A Mind at a Time." Levine explored this topic further in his book "The Myth of Laziness."

The MYTH of LAZINESS

America's Top Learning Expert Shows How Kids— and Parents—Can Become More Productive

MEL LEVINE, M.D.

Section Three: Getting the Data

Using Existing Data

Ethnographic Research and Usability Studies

15. Using Existing Data

The shrewd guess, the fertile hypothesis, the courageous leap to a tentative conclusion—these are the most valuable coin of the thinker at work. But in most schools guessing is heavily penalized and is associated somehow with laziness.

—Jerome S. Bruner

Sometimes, the target audience for a product is well-defined and shares certain cognitive strengths and weaknesses which need to be considered in the design. You can try to use cognitive characteristics defined by the Cognitive Wheel (Chapter 5) and existing research about a particular audience group to understand the needs of your users. The mantra for a product designer must be "Know Thy Audience."

Experts & Computers

> *An expert is someone who knows some of the worst mistakes that can be made in his subject, and how to avoid them.*
>
> —Werner Heisenberg

There are two types of experts when it comes to computer interaction. There are those individuals who are particularly facile with computers and don't feel intimidated by new computer-related tasks. These are the **computer experts**. And there are those people who don't know computers well but are experts in a particular domain of knowledge. These are the **domain experts**. A computer expert will be very good at mastering a computer interface but will need help to learn the subject matter. A domain expert will know a lot about the subject matter but might need help accessing the features of a computer program. Most people are a combination of both in any particular situation.

Domain experts tend to know the result they want, but don't know how to get it. Computer experts are just the opposite. They are comfortable with giving commands, but they are not sure what result they really want. Both of these difficulties lead to frustration and need to be accommodated differently.

In short, these two types of experts are distinctly different and each should have products designed for them accordingly.

Good design builds upon the base of knowledge already possessed by the prospective audience. If you know that your audience is composed of computer experts, then focus on presenting the underlying information clearly—they'll grasp the operational part of the software quickly. If the expertise of your audience is in the subject matter underlying the software, then focus on making it easy to access the software's features. Your goal, then, is to create an interface that connects concepts and actions that this audience already knows with the tools available in this new computer application.

Often interface metaphors are chosen to attempt to link the expertise that the audience already has to the feature-set of the software program. There are a host of audio editing and manipulation programs that look like on-screen representations of professional audio equipment, down to the 3D rendered sliders with specular reflections off the corners. I will discuss the use of these types of metaphors in Chapter 19: "Design Recommendations."

Upgrade Path for Computer Experts

Technology is like fish. The longer it stays on the shelf, the less desirable it becomes.

—Andrew Heller, IBM

An interface should always consider the needs of those users who have mastered its commands and now need to use the tool while being hampered as little as possible by features designed to assist novices. An example are the keyboard shortcuts available for expert users in many programs who have outgrown slower graphical point-and-click commands.

In general, we like to design interfaces that contain scaffolds which help to support novice users but which get out of the way of more expert ones. The most elegant scaffolds are those which are invisible, but sometimes it is appropriate to allow them to be turned on and off as a customizable preference.

One can imagine a program which teaches piano composition. Perhaps, as the user plays a keyboard attached to the computer, his key presses are forced to a strict underlying rhythm. Perhaps little lights flash on the next key to be pressed. Perhaps a metronome clicks during recording. Perhaps only notes in a certain key are accepted. These features might all be very valuable to a novice, but a user needs to be able to switch them off. These features, helpful in the beginning, shouldn't be allowed to become nuisances as the user's skill increases.

Another approach to creating an upgrade path for expert users is to limit novices to a subset of the functionality of a piece of software. I've seen this done either as an option or by compulsion.

HSC Software, in their KPT Convolver product, forced novices to use a simplified interface until they had demonstrated their competence to the satisfaction of the software itself by accessing certain features or spending a certain amount of time doing some action. The software then rewarded the user by turning on some additional related functionality. While this is an interesting approach, I personally found it too annoying and almost patronizing to be effective in a tool-type application. Beyond the negative emotional reaction that a user might have in being denied the full functionality of the software that she's bought, the designer had to be aware that people frequently reinstall a piece of software for a hardware upgrade or other maintenance reasons. An expert user will be very dissatisfied with software which refuses them access to functions they

know exist because the software hasn't yet acknowledged their expertise again! In a piece of entertainment software, however, this kind of approach is often used as part of game play.

Kai Krause's interfaces—he's the creator of KPT Convolver—while frequently stunning and always innovative, can be criticized primarily in that they do little to harness the expertise that a user might have in other software applications. His philosophy seems to be that users will approach his products as if they were games or pieces of art, exploring and touching all the little buttons and switches and then admiring the interesting effects. But his tools are often just one of a slew of tools that a graphic designer uses in trying to achieve an effect. Kai's interfaces generally require that the user learns each one from scratch.

Dumb as a Doornail

Education is what survives when what has been learned has been forgotten.

—B. F. Skinner

Where is all the knowledge we lost with information?

—T.S. Elliot

A product designer needs to understand her audience, her user population. Sometimes, on the road to defining that user population, it helps to consider worst and best case scenarios. What are the characteristics of the smartest, most flexible, and easily adaptable users? Who are the "dumb as a doornail" users and what do they need to succeed in using your product? Unfortunately just creating a taxonomy of users by dividing them into smart and dumb is not very useful—smart users will always figure out how to use the product, and dumb ones never will. As a product designer, where can you go from there? And by putting a "stupid" label on a user, it becomes difficult to be sympathetic to that user's plight: if a dumb user makes a mistake, it's his fault—the designer is suddenly relieved from the responsibility of making reasonable accommodations for all the users of the product. So it's not enough to define the two user extremes. The designer needs to understand why the failures happen and how to turn those failures into successes.

In educational research literature, there is a cognitive characteristic which generally refers to an individual's ability to learn and understand new information: **Degree of Field Dependence**. The concept of Field Dependence was first introduced by Herbert Witkin in 1948 and has been a topic of educational dissertation research for dozens of scholars. This cognitive category, as traditionally defined, is a large catch-all for cognitive "stuff" and is briefly described below. While I believe that the Cognitive Wheel described in Chapter 5 is a good tool for understanding prospective audience for a product, the Degree of Field Dependence has been the source of many experiments in the last 50 years and numerous interventions have been developed based on this cognitive characteristic. It's instructive to see how curriculum instruction—an education product—can be adapted to the needs of a particular audience based on this characteristic (which strongly resembles the two user extremes: smart and "dumb as a doornail").

Field Dependence

In the educational research literature, the definitions for **Field Dependence** and **Field Independence** tends to encompass a broad range of differing cognitive characteristics, changing from author to author. Here, I've attempted to provide a more narrow definition of Field Dependence to make it more meaningful and distinct from other cognitive variables discussed previously. The research background section provides the theoretical background and reasoning for the chosen definition.

Field Dependence and Field Independence are two extremes along a continuum. The degree of Field Independence or Field Dependence is based on the individual's ability to distinguish relevant information from surrounding perceptual and conceptual noise.

Field Dependence is a characteristic which, at its extreme, implies a low tolerance for ambiguity, cautiousness, low formal operational reasoning, and high dogmatism. Field Dependent individuals tend to have a high dependence on teachers and authority figures and desire less freedom and control of their environment. They prefer authorities to tell them what to do. They have little intellectual curiosity and tend to learn only what is required of them. They also have a tendency to view things in absolutes: strong versus weak, good versus evil.

Field Independence is a characteristic which, at its extreme, implies a high tolerance for ambiguity, intellectual risk taking, high formal operational reasoning, and low dogmatism. Field Independent individuals tend to desire greater freedom and control of their environment. They have less dependence on authority figures. They have a high degree of intellectual curiosity. They also tend to be flexible learners who are not easily distracted from their tasks and are better able to inhibit irrelevant response. They prefer sensory-rich stimuli and complex concepts that require higher order thinking skills. They enjoy the theoretical and philosophical underpinning of the material they are interested in learning. They also tend to develop complex, multidimensional descriptions of mental constructs.

It is not desirable to be Field Dependent, and it is desirable to be Field Independent. The good news is that this characteristic is not genetic and changes over time. In fact, the more educated you are, the more Field Independent you tend to become. This seems obvious since the more time you spend in school, the more time you have to learn how to learn.

Field Dependence Research Background

The original identification of Field Dependence and Field Independence is credited to Herman Witkin, a social psychologist, in the 1940s. Witkin used two tests to study individuals' abilities to orient themselves in space while immersed in a confusing perceptual field.

The first test was the **Body Adjustment Test**. A person was placed in a tilted chair, which in turn was placed in a tilted room. The task was to orient the body along the absolute vertical axis. According to this test, a Field Independent person could align their body along the absolute vertical axis, and a Field Dependent individual had trouble doing so.

The second test was the **Rod-and-Frame Test**. During this test, a person was placed in a dark room with a tilted luminous frame containing a tilted luminous rod. The job was to align a luminous rod along an absolute vertical. Again, a Field Independent person could easily perform this task, while a Field Dependent individual could not.

The terms Field Dependence and Field Independence comes from these tests—a person who could perform the orienting tasks in these tests was labeled independent of the surrounding field of information. The surprising outcome is the correlation between the individual's performance on these tests and his ability to learn and do well in educational settings.

Since then, a lot of research has been done on the effects of Field Dependence and Field Independence on learning outcomes. And research-supported lists of deficits and strength have been developed for this cognitive characteristic. For more information about Field Dependence research see Appendix.

Field Dependence and Product Design

> *We live in a society exquisitely dependent on science and technology, in which hardly anyone knows anything about science and technology.*
>
> —Carl Sagan

There are two approaches to developing content that should be used in tandem: working with an individual's strengths and providing scaffolding to compensate for weaknesses. In the long term, both are equally important. But some online situations (e.g. just in time learning) dictate the design approach be one way or the other. Either way, it helps to understand what an individual's strengths and weaknesses are.

Field Dependence and Field Independence is defined in the Cognitive Wheel through a set of qualities an individual will exhibit in a particular setting: background knowledge, attention controls, etc. General design strategies for Field Dependent individuals are:

- design a very structured environment

- provide continuous feedback on performance in multiple media (e.g. auditory cues, animations, change of color, text messages, video, etc.)

- develop a deductive presentation sequence

- serve small informational objects (short sequences)

- include multiple organizers of information (e.g. maps, table of contents, outlines, etc.)

- generate low stress environment (e.g. a cooperative rather than competitive environment, group work)

- create very clear directions and rules, giving maximum amount of guidance and support to do an activity

- identify key ideas, rules, concepts, principles, and definitions

- model appropriate behavior

- give lots of examples (e.g. prototypical and distant examples, divergent and counter examples, personal examples, etc.)

Given the nature of the deficits of people who are Field Dependent, a large collection of design tools is needed to support them.

In a context of online learning, these instructional strategies need to be translated into specific **Interface and Content Prescriptions** (a list is provided in the Appendix). Using that structure, those parameters can be set as follows:

- Instructional Sequence: deductive, easy to difficult sequence in form of presentation

- Community Content Type: moderated

- Level of Interactivity: low to medium

- Level of Participation in a Community: low to medium

- Duration of Interaction: short

- Emotional Impact: positive

- Organizational Schemes: multiple

- Control over content, navigation, interface, etc.: limited

Field Dependence Assessment Possibilities

There are multiple tests that assess an individual's degree of Field Dependence. A lot of these tests ask a person to find a hidden simple shape inside a complicated illustration. Some tests ask to identify a simple tune inside an elaborately orchestrated piece of music. All of these tests are timed.

In the case of online learning, a simple variant of an embedded shape test can be used as a pretest in an online portal. Once tested, the portal can apply differentiated design to personalize the instruction.

Since the degree of Field Dependence changes over time, it's important to keep track of individual's performance and preferences—what was once useful could easily become annoying.

Seniors

Education is the best provision for old age.

—Aristotle (384-322 BC)

While it might sound obvious, there are a set of changes that we undergo, both cognitive and physical, as we age. Unfortunately for seniors, most products are designed by youngsters. For a while, this was a business strategy for technology companies—grandparents tend not to be the early adaptors of technology; rather, they inherit old stuff from their grandchildren. Fortunately, this vision of old age is changing and this demographic is acquiring a great deal of influence over product design.

In Japan, where the population of seniors outnumber any other age group, businesses are springing up that cater exclusively to this group of consumers. There are supermarkets that widened their aisles and lowered their shelves to better accommodate patrons with walkers and wheelchairs. Produce is sold in smaller quantities to better meet the needs of a person living alone. Signs have increased in size to be seen better by those with weak eyesight. These markets also introduced benches and shopping carts with seats. Even the type of produce sold caters to the taste of seniors. These and many other subtle changes make these shops more attractive to an older audience.

The adaptations which product designers have to make to gear their creations towards an older user group can be broadly classified into **physiological** (e.g. mobility issues, poor stamina, etc.), **perceptual** (e.g. failing eyesight, bad hearing, etc.), **cognitive** (e.g. memory problems, attention control problems, etc.), and **content focus** (e.g. news and health information aimed at seniors, fashion and food items for the needs of an older audience, etc.).

Take poor eyesight for example. Most people in their fifties experience the onset of far-sightedness. This condition renders an individual incapable of reading small text at short distances. A direct prescription for product designers is to use large text and use it consistently throughout. Dark text is easier to read on light backgrounds, so the use of black for body text on a white background is generally a good idea. Another way of dealing with poor eyesight is to use other navigational and informational devices to signal the user when there has been a change in content or a change in activity. The simplest such device is the use of large blocks of color to help navigate the user through the content or space. While body text is black on white, section or area heading text can be placed against color—these usually don't change often and so they will be read only a few times. The user will then identify the content area primarily by color and placement. Color could follow the heading to become the page topper color for the secondary pages of a Web site, for example. This is a very effective navigational device and useful not only for a senior audience. In fact, many designs which start as aids to accessibility for disabled turn out to have a broader utility.

As people grow older, it becomes harder to commit information to long-term memory. Short term memory and working memory especially also decline with age. The direct outcome of this problem is that older individuals tend to get lost easily in navigational structures. Confusion about one's location leads to frustration and dissatisfaction. There is no simple fix to this problem except to be especially careful to create well-designed navigational schemata. Navigational tools should be consistent from page to page and location to location, and their placement should be identically situated—a user should never have to think twice about where to find navigational information.

There is a higher incidence of mobility problems due to strokes and other ailments in a population of seniors. They might have more difficulties using mice or keyboards, for example. Keyboards designed for one-handed input, a larger mouse or an alternative input device, and larger clickable areas on the screen are all reasonable ways to accommodate this problem for computer-based products.

My company once designed an interface for an audience of stroke victims. Stoke victims have a host of disabilities, often including hemispherical visual field inattention—these patients only really see either the right or left side of the visual field. It was important to avoid placing all navigational items on one side of the computer screen or the other because a number of the users wouldn't be able to find the controls. So we discreetly duplicated navigational elements on both sides of the computer screen.

Take a look at these Web sites geared to an older audience. Do these sites adequately deal with the needs of their audience?

> http://www.ThirdAge.com
>
> http://www.SeniorNet.com/

Kids

Kids are often lumped into one group as if they shared an amorphous characteristic of "kidness" which can be targeted by a software product. But there is no reason not to assess children using the same cognitive traits described in previous sections. Important characteristics to watch out for are Background Knowledge and Meta Knowledge—the Cognitive Wheel categories that are directly related to how old the children are and what have they had time to learn and absorb.

To start, for product designers, the most obvious and tangible developmental difference between children ages 4 to 7 and those aged 8 to 12 is that the former group does not have fully developed reading skills. On average, reading proficiency stops being a limitation after about 8 years of age, when children learn to read fluently. While younger children often learn the alphabet and are able to recognize their names when written, those skills are based on pattern recognition rather than an actual ability to read. Children are judged able to read when they are capable of understanding that a set of letters in a particular sequence makes a word that has actual meaning. Children who are able to read can read words that they've never seen before.

It's helpful to understand just how it feels when you have very few clues from a written language. English-only speakers might find it instructive to check out this Web site:

> http://www.willemwever.nl

Piaget's Stages of Development

A teacher who can arouse a feeling for one single good action, for one single good poem, accomplishes more than he who fills our memory with rows on rows of natural objects classified by name and form.

—Johann Wolfgang Von Goethe

Jean Piaget did extensive research on children's understanding and cognitive development. His insights provide valuable clues to the workings of a child's mind.

Piaget identified three major developmental stages in children. They were the **Sensorimotor**

Stage (birth to 2 years), the **Preoperational Stage** (2 to 6 years), and the **Concrete Operation Stage** (6 to 12 years). Each of these stages are, in turn, divided into substages that are characterized by children's abilities to perform a series of ever more complex tasks. According to Piaget, the ability to perform these tasks is fundamentally limited by the development of their working memory and by experience.

As children develop from age 3 to 11, they go through a number of qualitatively distinct substages in which their thinking becomes progressively more systematic and logical. Piaget identified these substages by the types of strategies that children use in solving problems. Children aged 3 to 4½ are only capable of using **Isolated Centration Strategy** to solve a problem. They succeed in identifying the presence or lack of an object—something is either there or it is not. Children aged 4½ to 6 move to the **Unidimensional Comparison Strategy**—their strategy becomes comparative rather than absolute. A child at this substage of development is able to notice the difference in the number of objects between two groups—one group has more and the other has less. (Dogs and other mammals can do this as well up to about 5 objects.)

By the age of 7, children begin to not be confounded when an additional element is added to the problem. They are said to use a **Bidimensional Comparison Strategy**. These children will be able to solve a problem where, for example, they are given two groups of marbles, each group containing a varying number of light and dark marbles. Children using a Bidimensional Comparison Strategy will be able to tell which group has more dark marbles, even if that group has fewer total marbles. Children using a Unidimensional Comparison Strategy will be unable to see that a group with fewer total marbles contains more darker ones.

However, if children who use a Bidimensional Comparison Strategy attempt to state which group has a higher ratio of dark to light, they will simply guess. If the problem is changed to include the size of the dark marbles as a factor, they will be unable to state which group has more dark marbles in it. Children aged 9 to 10 are said to use **Bidimensional Comparison with Quantification**. They would be able to solve the foregoing problem of the number of dark marbles irrespective of size, although if asked to specify which group has a higher ratio of dark to light marbles, they will also probably guess.

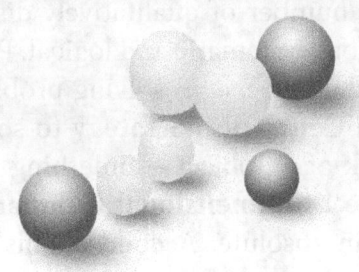

As can be seen from the marble examples, at each developmental stage, the number of variables presented to a child increases. First, there was only one thing to pay attention to: is the marble there or not? Next, a child can compare two groups of marbles—which has more? At the third stage, the comparison problem becomes more difficult by the addition of another variable—color. Which group has more dark colored marbles? An addition of a size variable increases the difficulty of the comparison once again: which group has more small dark marbles? These are nested comparison problems, each more difficult than the one which came before it.

Clearly, if you train a child to solve these kinds of marble problems, you might get a much younger child to figure out a multi variable comparison. But would this knowledge translate into other domains?

Piaget's stages of development do not imply that all kids go from one stage to the next at exactly the same age. Rather, it's the sequence of the stages that is invariant—a child can't get to the Concrete Operation Stage without first passing through the Preoperational Stage.

This is all relevant to assisting a product designer in understanding the mentality of the target audience. A child under 8 will consistently get lost in a hierarchical structure, for example, but will do well in a directional one. A **geographic interface** provides more navigational clues and reduces some of the demands upon their working memory. Kids under 8 do well in a Living Books type environment and poorly in an adventure environment—one in which they are asked to take an object from one location and use it somewhere else not associated with where they found it.

While we admire these products, both "Pajama Sam" by Humongous Entertainment and "Menlo the Frog" by Windy Hill Productions are unplayable by the younger preschool members of their targeted audience range (3 to 8) if these kids are not given initial assistance through the game. With repeated exposure to the games, however, even younger kids will eventually be able to play these games unassisted.

Beta testing with young kids is valuable, but care should be taken to not only select children with extensive computer experience. Kids with a lot of computer experience develop coping strategies that make them function better in interface situations that haven't taken account of their development abilities. For example, a young child may have learned a strategy of clicking everywhere, which often works but doesn't demonstrate the efficacy of an interface design. Even the youngest computer users sometimes achieve real computer competency and understanding, but they are in the minority.

Detail Over Function

Another developmental difference between kids roughly above and below nine years of age is **Gardner's Literal Phase**. Younger children tend to spend their creative efforts on the fine details of their creations. So if they are drawing a dinosaur, their attention and interest is taken by the details of the color of the skin, the shape of the horns, the size of the teeth. Little attention is paid to the function and the underlying structure of the different parts of the animal. So, for example, in games like SimCity, the younger kids are much more interested in adding building to their cities, while the civil engineering concerns of the city are ignored.

The younger kids are more concerned with what they can actually see. If it's there, they can describe it. But changes to an object over time, such as the behavior of a simulated city population, are much harder to recognize. Older kids and adults can do it. But even then, the subtleties of recognizing evolution over time is hard. And the longer it takes for something to change, the harder it is to notice. So in SimCity example, a younger child will concentrate on what he can easily see and understand—the layout of the city. He wants the layout to be pretty. If a gas station looks ugly, then out it goes.

One way to overcome the above limitation is through repeated experience. If a young child does an action a couple of hundred times, after awhile they'll expect it. More than that, they'll demand it. The younger the child, the more resistant he is to changes in his regime. This applies to games as well. Once a young child learns to do something a certain way, he'll be very uncomfortable with changing that procedure.

Computer-based Products for Kids

The last good thing written in C++ was the Pachelbel Canon.

—Jerry Olson

What does this all mean to product design for kids? The first rule of thumb is to build on what children already know. An experienced young computer user will have had exposure to certain basic actions that are present in most children's software packages. So familiarity with popular kid's software is an important start. Our children expect that programs can be saved in mid-game and retrieved at a later time. The classic way this is done is by taking a snapshot of the screen at the moment the child wants to save the game and then using that snapshot as an icon for the saved game. The kids should also be allowed to type in their names under the saved game icon, but it shouldn't be necessary.

Another serious difference between designing for kids rather than adults is the mismatch in the goals of the children who are using the product as opposed to the goals of the designers. The basic dichotomy is education versus fun: adult goals versus kids goals. Please see the section entitled "Shaping of Goals" in Chapter 19 for an example involving kids.

Personality Test for Kids

I have not failed. I've just found 10,000 ways that won't work.

—Thomas Edison

Kids develop personality traits at a very early age (some say from before birth). Using the available research from David Keirsey, Isabel Myers, and Kathrin Briggs, I constructed a quick guide to sorting kids into different personality groups. While this method of sorting kids is not as reliable as the adults version where the adults get to decide for him or herself the best possible fit for each question (for kids it is one step removed from the source of information), it still can provide valuable data. And just like the adult groups, for kids each personality group has interface and content prescriptions associated with it. Some become obvious from reading the descriptions. Some can be taken from adult product interface and content prescriptions and adapted appropriately. The portraits of the four main personality combination types are as applicable to kids as they are to adults once modifications are made for background knowledge and other cognitive developmental constraints.

For each of the following, please appreciate that we all have a bit of each extreme. But, for the most part, we will tend to operate in our own preferred mode, which may lean towards one pole or the other. Your task is to think of a child that you know and check the observations that describe that child in a majority of situations. Count the number of A's and B's in your answers to each of the categories.

Expressive vs. Reserved

If there are more A's in your answers to this section, then the child probably leans towards an Expressive personality type.

A. The child is quick to act and to approach new people.

B. The child is shy, quiet, and unobtrusive.

A. The child jumps into the conversation and is quick to verbalize.

B. The child holds back responses until he or she has rehearsed internally.

A. The child is ready to enter into group activities, and finds and makes friends quickly and easily.

B. The child exhibits slow development of social skills.

A. The child is usually in line with public opinion and tends to be on the side of majority.

B. The child tends to be judged "stubborn."

A. The child is responsive, expressive, and enthusiastic.

B. The child is reserved, needs privacy, and is hesitant to share personal ideas.

On average, in a group of four, only one will be Reserved (Introverted) and the other three will be Expressive. And as a society, we reward the quality of "Expressiveness." Expressiveness and Reservedness might be rewarded differently in other cultures, who might find silence, for example, a more desirable personality attribute for children.

Observant vs. Introspective

If there are more A's in your answers to this section, then the child probably leans towards an Observant personality type.

A. The child prefers new stories with lots of details.

A. The child likes sequential adventure type stories about the real (factual, familiar) world.

B. The child asks for repetition of stories and wants the same book read over and over again.

B. The child prefers fantasies with knights and dragons, kings and queens.

A. The child prefers playing a game to story time.

B. The child daydreams for hours.

A. The child shines in a world of action. This doesn't necessarily imply sports activities although it could. It could also imply construction play or hands-on activities.

B. The child is very "opinionated"—knows the "right" way of doing things.

A. The child's motivations are easy to understand.

B. The child is "his own person"—he does things his own way.

A. The child enjoys detailed work like coloring books and doing workbooks.

B. A child's toy truck is easily transformed into a rocket or a submarine.

Like the Expressive vs. Reserved category above, on average, in a group of four there are three Observant individuals and one Introspective person.

Reasoning vs. Emotional

If there are more A's in your answers to this section, then the child probably leans towards a Reasoning personality type.

A. The child needs and asks for reasons behind an activity or a request: "But WHY do I have to do this?"

B. The child will perform small services for parents and teachers in return for expressed gratitude. The child aims to please.

A. The child will emotionally detach him or herself from an unfavorable emotional circumstance.

B. The child is very sensitive to the emotional climate at home and can get physically ill during a conflict.

A. The child is not very interested in interpersonal dynamics and tries to seek out objective facts.

B. The child enjoys involvement in family, school, and neighborhood happenings (gossip).

A. While appearing indifferent, the child takes in emotion very deeply.

A. The child doesn't display emotions expressively and tries to block facial expressions.

B. The child cries easily.

A. The child tends to be unaware of the emotional distress of those around him or her.

B. The child tends to be aware of emotional and physical comfort of others.

There are as many Reasoning individuals as there are Emotional ones.

Schedulers vs. Probers

If there are more A's in your answers to this section, then the child probably leans towards a Scheduling personality type.

A. The child wants order.

B. The child seems indifferent to established ways of doing things.

A. The child gets ready on time and worries about being late.

B. The child is unconcerned with temporal constraints—a deadline is only a suggestion.

A. The child seems very sure of him or herself.

B. The child is more tentative in speech patterns and qualifies statements.

A. The child likes his or her stuff (toys, books, clothes, room, etc.) in neat order.

B. The child's room is a jumbled mess.

A. The child asks for and initiates daily routines

B. The child may have to be reminded to eat and get dressed, etc.

There are as many Schedulers as there are Probers—they are equally distributed in a population.

Additional Thoughts and Further Readings

David Jonassen's and Barbara Grabowski's textbook on psychometrics, "Handbook of Individual Differences, Learning & Instruction," is a great place to read more about Field Dependence. See also Anita Woolfolk's "Educational Psychology." And David Keirsey's and Marilyn Bates' book, "Please Understand Me: Character & Temperament Type," provides information about children's personalities.

16. Ethnographic Research and Usability Studies

Things are not always what they seem.

— Phaedrus

Ethnographic Research

Ethnographic research allows product and space designers to learn and understand how individuals actually interact with objects and locations through close field observation of sociocultural phenomena. Typically, a product designer focuses on particular practices that are involved in a very narrowly-defined set of interactions: people purchasing tickets for mass transit rides; individuals using ATMs, etc. Designers interested in developing larger projects—like museum exhibits, supermarket layouts, or train station organization—focus on studying people's reactions in those spaces. In both cases, designers need to identify the likely user groups—what are their defining characteristics—and their goals for interacting with the product.

Ethnographic research is done via field observations over extended periods of time and interviews with individuals—informants. Field notes, photographs, video tapes, and questionnaires are the tools of ethnographic research. Data analysis and theory formation come at the end of the observation period. As an ethnographer, you spend a lot of time observing different people doing the various tasks of interest, at various times of days, among a particular segment of the population—your user interest group. Your focus of research might be as narrow as a particular interaction or as wide as all of activities

Ethnographic Research Tools:

- **Field Notes**

- **Photographs**

- **Video Tapes**

- **Questionnaires**

for a set duration of observation.

Product and space designers focus on micro-ethnography as opposed to macro-ethnography. **Micro-ethnography** is the study of narrowly-defined cultural groupings, such as "English speaking tourists using Paris buses" or "foreign students studying in Paris." In both of these cases, it's important to further delineate the differences and similarities among members of these groups, e.g. foreign students from Australia versus foreign students from England; US tourists from California versus US tourist from Mississippi; tour groups versus family groups visiting the same museum; foreign students fluent in French versus foreign students with limited French language skills.

Product and space designers also take on the "Etic" perspective (rather than the "emic" perspective). **Etic** perspective is the ethnographic research approach to the way non-members (outsiders) perceive and interpret behaviors and phenomena associated with a given situation or, more broadly, a culture. **Emic** perspective is the ethnographic research approach to the way the members of the given culture perceive their world. The emic perspective is usually the main focus of ethnography.

Etic micro-ethnographic research questions focus on the causal relationship between behavior, understanding, and social knowledge of a particular cultural sub-group. The data consists of event descriptions (via field notes) and interview answers from actual participants or informants.

Not Everyone Can Invent a Wheel

> *How odd it is that anyone should not see that all observation*
> *must be for or against some view if it is to be of any service!*
>
> —Charles Darwin

There is a wonderful episode of "The Simpsons" which features Homer reuniting with his long lost multimillionaire brother who asks him to design a car for his automobile manufacturing business. Homer, being an expert car driver—he owns one and takes it to work every day, after all—should be an ideal car designer for an "average" guy. Homer wants bigger seats, an extra large cup holder, and a few other modifications that would make the car extra cool and drool worthy. But unfortunately (as I'm sure you've guessed by now), Homer's designs prove too expensive and impractical, and the factory, overburdened with Homer's creativity, goes bankrupt.

So what went wrong? Why couldn't an average expert driver design a car for his user group? Or why couldn't a hair stylist develop a new hair drier? We all use toilets, wouldn't that make us ideal designers of one? Clearly, having expertise in using a product doesn't immediately translate into the engineering and design chaps necessary to create one.

In the dawn of time, I imagine each person knew all there was to know about survival: how to make a bed nest in a tree; how to start a fire; what roots were good to eat and which were poisonous; and so on. Those that couldn't figure it out didn't live long. But at some point, being an expert at everything became impractical. There were farmers and blacksmiths, and farmers didn't have to learn the metal trade to get metal tools—they traded food for tools, and everyone was happy. This didn't mean that farmers couldn't inform the design of a scythe, for instance. But an average farmer could only judge if a tool felt good and comfortable while he worked in the field. It took an extraordinary farmer to understand how to make improvements to his tools and how to get the blacksmiths to execute his designs in metal.

Design and Usability Testing

> *I think there is a world market for maybe five computers.*
>
> —IBM Chairman Thomas Watson, 1943

So if the user can't be asked to design a product, what can she be asked to do? Products are designed to fill the need, expressed or imagined, of a specific audience. For example, prior to designing a new automobile for a family market, car makers have to combine what they know about car design with the market research on what the users from that group of buyers want. Before any sketches hit the table, manufacturers try to guesstimate what this car would be used for and by whom—this is the realm of conceptual design. Below are some of the myriad design questions that would be considered at this point.

- What is the price range that this user group can afford? (This is a very important question to determine feature development.)

- How many seats does an average family need in a car? There is an average family size, but school carpooling practices need to be considered also.

- How big of a baggage compartment do families need? (Are they COSTCO shoppers? Do these people camp?)

- What is the acceptable gas mileage?

- How important are the safety features? (Safety features are clearly very important, but how much are consumers willing to pay for them?)

- How long does the family tend to own a car? Do the car seats need to be built-in or be removable, or do families upgrade cars more often than the five years of practical car seat use?

- Would removable seats be a desirable feature? Would consumers pay for them?

- How important is on-board entertainment? (The new fad is built-in TVs with DVDs and game consoles for riders on the rear.)

- How big are average family garages? (These cars need to fit.)

- Are there special regulations that need to be accommodated for each state? (California has strict emission control laws, for example.)

- What car features have become a necessity and which are still considered luxury items? (Air bags used to be luxury items, but now are standard.)

This background research will inform the conceptual design of the family car and can be conducted via market trends research—what car buyers are actually purchasing—and by questioning the prospective buyers—what buyers would like to see in a car.

The next step is interaction design: how to achieve the conceptual designers' vision for this family car. Among the many thousands of details, these will include:

- How wide should the seats be?

- How far off the ground can an average driver step up to get into the car?

- How big and thick should the steering wheel be?

- Where to put all those CD and radio controls and how to deal with mode errors not to mention the whole slew of design issues that relate to the simultaneous driving and operating of entertainment units in the car?

- How big should the mirrors be and where?
- How does the sound travel in a car and how to foster conversation and good sound acoustics?

There are so many things to consider during interaction design phase. And these design decisions are not made by a single person but by multiple teams, each member of which has fairly narrow specialization—some might just specialize in the design of car speakers, for example. To get information on whether interaction design is successful, potential future users have to be tested. The average ergonomic factors for these drivers can be obtained through such user tests or from data that has previously been gathered. Drivers can be monitored during typical car outings to see how much of their working memory is being allocated to driving as opposed to other activities they do in the car—imagine taking five kids to a baseball practice or talking on the cell phone. The safety issues can be tested with crash dummies. Drivers can be observed for the type of errors they are likely to make and under which circumstances they are likely to make them. Cognitive scaffoldings could then be integrated into the design to support users in situations of failure—perhaps a voice activated driver assistant that talks the user through turning on the radio or adding air to tires.

The interface of the car deals with how these different functions look and feel. This would include designing supports for perceptual processing and attention controls. For example:

- Can drivers easily tell when they are in reverse prior to crashing the car?
- Is it clear when the headlights are on?
- Is it easy to see which radio station is being listened to?
- Is there an indication that the back door is open?
- Do the seats feel sticky during hot weather?

Clearly, conceptual design, interaction design, and interface design are closely interconnected—it seems impossible to just focus on one without having some understanding on the impact of design decisions for all.

Once the prototype of the family car is constructed, more user testing can be done with an actual vehicle. Drivers can be asked to maneuver a difficult course or do parallel parking—a list of these benchmarks would be made prior to beginning the user testing. User testing can be done with professional drivers and with average users. A set of questions can be designed to elicit users' feelings and experiences with the new car:

- Did you feel comfortable making that U-turn?
- Was it easy to put on the seat belt?
- Could you see well in front of you?
- Are the mirrors easy to adjust?
- Does this car do all that you need it to do?

While it's okay to ask what the users would want to do with their cars (e.g. are they likely to be transporting a canoe in their new car), it is the designer's job to figure out how to accommodate those needs and desires (e.g. create roof-rack or allow the removal of seats or a drop-down back window or some other solution to transporting a canoe with the family car).

You should never ask the users to design the product. But you do want to elicit information about users' opinions: Do they like it? Are they comfortable? And most importantly, would they buy the car?

So you must be clear about what you can expect from a focus group or what user testing will reveal. Focus groups are useful for identifying major flaws in the design of a product. But you can't expect a focus group to redesign your product for you. Asking a focus group how they would fix a particular identified issue is a mistake. Allow the focus group to point out issues that are within their expertise as users and then allow designers to come up with solutions.

Remember that you only get out of a focus group what you put into it—the more leading the questions you ask, the more inaccuracies you introduce into this process. And the results of a one-day focus group session can never be of any statistical value—**the sample is too small**. The next time someone boasts that four out of five users feel a certain way, ask how many users they've asked—was it five?

Assessment Tools

Remember: you are the only person who thinks in your mind!
You are the power and authority in your world.

—Louise Hay

There are two types of assessment tools available to product designers: summative and continuous. **Continuous assessment** is a dialogue with users as they interact with the product or a continuous measurement of user performance during the interaction (e.g. eye tracking).

Summative assessment is assessment after the fact. It could be a set of questions that the users are asked *after* they have had a chance to interact with a product (like a final exam). Or it could be a set of tasks that the users are asked to perform and which are then scored, using a target performance as a base line.

Usability Script

The more prepared you are, the better are the results of your usability testing. Optimally, you come into a usability lab with a script—this is a continuous assessment tool. All test subjects get tested with the same script, which is preferably administered by the same moderator. This approach gives a certain amount of uniformity to the test data.

The usability script is designed to test conceptual, interaction, and interface design flaws. It is a collection of ever more precise prompts that guide the user through the product. Consider a Web design usability test. If the usability subject is sitting and looking at the computer screen for longer than you expected her to do so—you have to define these time periods in advance—you might offer a bit of first level help:

Conceptual design prompt is the first level prompt:

Do you understand what to do?

This is a conceptual design question—it deals with what the product is designed to do. If the user doesn't get this, no amount of pretty buttons will solve the product design failure.

The next level of help deals with interaction design:

You can do this by using that tool or by going to that page or going to that space.

If the user understands what she needs to do but can't figure out how to do it, it's an interaction design failure.

And finally, you get to test the interface design:

Press this type of button.

This is the final level of help—the user understands what to do and how to do it, she just didn't see the button. This is where making a button bigger, or more colorful, or flashing, or better labeled can actually solve the design flaw.

A Quick Summary of Usability Factors

640K ought to be enough for anybody.

—Microsoft Chairman Bill Gates, 1981

There are three factors that can be measured by usability studies: **usefulness**, **learnability**, and **satisfaction**. Each of these can be broken down into the conceptual design, interaction design, and interface design components. A good usability script would unearth problems with any of these components.

Usefulness

Conceptual Design:

- Are the goals of the product designers and the product users the same?

- Does the product meet the expectations of its users?

- How useful is the product for the purposes for which it was developed?

Interaction Design:

- Can the product users use the product? Is it too difficult to use?

- Is the user's performance on a task enhanced by the product? Does the product assist the user in doing something faster, better, etc.?

Interface Design:

- Are all the functions of the product clearly visible to the user?

- Is the product too cumbersome to use? Is it too heavy? Is it too bulky?

Learnability

Conceptual Design:

- What is the minimum level of desired performance?

- How much training do users need to achieve the minimum level of desired performance?

- Does the learnability curve match the expected usage of the product? (See the section "Learning Curve" in Chapter 9.)

- How forgiving is the product to a user error? Are there enough "undo's" built into the system?

- Is there built-in help?

- Has the right metaphor been adopted for the product? Do the users get it?

- Have users' expertise with similar products been appropriately utilized? Have industry standards been followed? Is the product using a standard conceptual model like a WIMP interface (windows, icons, menus, and pointers), for example?

Interaction Design:

- Are there enough rewards built into the product for novice users?

- Are expert users burdened by the design features intended for novices? (See Chapter 15: "Using Existing Data.")

- Is the built-in help actually helpful?

- Do the features of the product rely on recall or recognition?

- Are there default settings built into the product? The product automatically fills in today's date, for example.

- How closely does the adopted metaphor conform to the use of the product? Do users understand where the metaphor breaks down?

- Have users' expertise with similar products been appropriately utilized? Have industry standards been followed? Do the product's windows and menus work like windows and menus of other products in the same field?

Interface Design:

- How sensitive is the feedback given to users during an error event? Error messages have to be useful and polite.

- How easy is it to get help? Is it contextually embedded in the product?

- How clear are the navigation items? Are the buttons labeled with users' expertise in mind?

- How well is the adopted metaphor executed? Do the icons bring to mind the intended object to the users, for example?

- Have the users' expertise with similar products been appropriately utilized? Have industry standards been followed? Do product's icons and buttons look and feel like icons and buttons of other products in the same field, for example?

Satisfaction Test

Conceptual Design:

- Would users invest the necessary time to learn the product?

- Do the users believe that the product has value to them?

- Does the product's value to the users match its price?

- Was the right metaphor adopted? Does it have the right emotional tone?

- How do the users feel about using this product?

- Would the users recommend this product to others?

- Is the product in "good taste?" Is it culturally appropriate for its intended audience?

Interaction Design:

- How do the users feel about themselves while they use the product? Do they feel stupid for not getting it?

Interface Design:

- Are the users offended by the colors or images used in the product?

- Does the look and feel alienate a potential user group? ("Oh, this product is for boys!" "I wouldn't use it—it's for old people.")

Clearly, there is a lot of crossover between these categories. Most of the data to assess usability can be gathered through usability scripts, but other techniques such as eye tracking can be used to evaluate the visibility of certain design elements. And product comparison studies can reveal issues with usefulness, learnability, and satisfaction levels of a product.

Additional Thoughts and Further Readings

Web Sites:

- www.FocusGroups.com is a directory of marketing research firms for research professionals.

- www.GreenBook.org is a worldwide directory of marketing research companies and services.

Books:

- "Developing Focus Group Research: Politics, Theory and Practice," edited by Rosaline Barbour and Jenny Kitzinger, 1999

- "Focus Group Research Handbook," by Holly Edmunds, 2000

- "Wilder Nonprofit Field Guide to Conducting Successful Focus Groups," by Judith Sharken Simon, 1999

Cost of Usability:

As of 2008, if a professional service is used, the cost of one focus group day is around $5,000. Of this sum, $2,000 to $2,500 is paid to a moderator (focus group leader). Another $2,000 is paid to the facility. And the rest, $1,000 to $1,500, is paid to the participants. The average fee for a participant is $50 to $75 for a one hour session. A teenager typically earns $40 per focus group. But a highly paid professional like a surgeon would get up to $500 per session. The typical number of focus group participant sessions per day is 12. But it's important to hire a few more participants so that "no shows" won't result in dead time for the focus group—you still have to pay for the moderator and the facility even if no one is working.

Section Four: Design Notes

Designing for Errors

Miscommunication

Product Design Recommendations

17. Designing for Errors

Oh God, how did I get into this room with all these weird people?

—Stewart Brand

Errors Due to Attention Controls Failures

Recently, I made a very stupid mistake. I had eggs boiling on the stove and the phone rang. It rang just as the timer for the eggs went off. With the receiver pressed firmly between my shoulder and ear, I grabbed the pot with the eggs and put it in a sink to dump out the hot water and dowse it with cold (a technique I learned from Julia Child to avoid that gray layer around the yoke). With the pot of eggs and water in the sink, I dutifully added soap to the mixture—that's what I usually do with a pot in the sink. Of course soap doesn't improve hard-boiled eggs and I was horrified at what I've done. I quickly washed out the soap from the pot and rinsed the eggs.

I was doing well with the eggs until the telephone call interrupted my activity and my train of thought—my attention shifted from making the eggs to talking on the phone, and I made a stupid mistake. While stupid, this error was certainly not coupled with dire consequences. At most, my eggs would have tasted bad, and I would have needed to make a new batch. But this particular type of attention control error has the potential to end very badly indeed.

Attention Control Errors:

- **Autopilot Errors**
- **Absentmindedness Errors**
- **Interruptus Errors**
- **Mode Errors**
- **Perceptual Blindness**

Autopilot Errors

The soap in the eggs example shows what can happen when we stop paying attention to what we're doing. This failure to monitor what we do happens mostly for things we are good at. Consider driving: you get into a car to drive to the store, but end up at work. The route to work has become automatic. We talk about "driving on autopilot." Sometimes, people describe this experience as suddenly noticing that they have arrived without a clear memory of getting there. If we fail to monitor our actions, we fail to keep our attention on what we are doing. So we get into the car with one intention, but at some point in the journey we start thinking of other things and fail to monitor what we are doing. **In these types of attention control errors the initial sequence is the same but the actions diverge some point, leading to a mistake.**

Autopilot errors occur when we are very comfortable doing something—we are experts at performing the sequence of actions. We rarely end up at a store when we are driving around looking for a particular address. So anytime a person is asked to perform the same action over and over again with only an occasional variation, there is a strong chance that autopilot errors will arise. Such errors are the bane of factory assembly line workers and can lead to serious injuries.

Imagine filling out a form and writing today's date in place of your birth date. We've all done this. In this situation, at the time you were writing the date, you have already "moved on" to the next item in the form. As you stopped paying attention to the writing, the small part of your working memory "desk" devoted to completing this task reverted to the sequence that's automatic: writing today's date. This is another example of an autopilot error.

Exam creators love to test attention controls, and autopilot errors are well represented in the multiple choice answers. For example, consider a simple division problem: What is the remainder after 26 is divided by 4? The numerator is not evenly divisible by the denominator resulting in a remainder. The multiple choice answers would include the remainder and the whole result. A student whose attention wonders off the work, could easily do the math right and get the wrong answer on this question. Autopilot errors are common in children with weak attention controls.

26 ÷ 4 = 6 and remainder 2

26 — numerator

4 — denominator or divisor

2 — remainder

6 — largest possible multiple of 4 which goes into 26

Absentmindedness Errors

Have you ever caught yourself putting your reading glasses in the refrigerator instead of the milk? Or throwing away the present instead of the wrapping paper? In both cases, you knew what you had to do—put the milk in the refrigerator or clean up the mess—but you just stopped paying attention to what you were trying to do. **The result was that you performed the right action on the wrong object.** This is an example of an **absentmindedness error**.

There are individuals who excel at absentmindedness. If you talk to them, they know exactly what they need to do, but somehow it always turns out wrong. They put on pajamas when they should be getting dressed for work. They pack the wrong book into their bags, not because they don't know what to take, but rather because they stopped paying attention to what they were doing. They work on the wrong assignments or save papers in the wrong file. These peoples' actions are not intentional and thus are not remediable by punishment or even embarrassment. It's the job of a product designer to save these people from themselves!

Interruptus Errors

It is likely that unlikely things will happen.

—Aristotle

An **interruptus error** occurs when you forget what was being done in the middle of an action. You opened the refrigerator but then forget why you were there—it just slipped your mind. You were trying to make a point, but forgot what you were going to say—you lost your train of thought. In both cases, working memory got overloaded with other thoughts or outside stimuli and something just "slid off the desk," resulting in an awkward moment. **These are also examples of attention control errors—failure to keep the ideas on the table by paying attention to them.**

People who daydream a lot or who get easily preoccupied or distracted suffer interruptus errors all the time. The same is true for anxious individuals—most of their attention is taken up by worrying, leaving only a small sliver of working memory space to deal with current reality.

Interruptus errors can be particularly damaging for individuals who have a smaller than average working memory. These people already have a hard time managing the in and outflow of information. If, on top of this limitation, their attention easily wanders away from the task at hand, their performance suffers.

Mode Errors

How could this be a problem in a country where we have Intel and Microsoft?

—Al Gore on Y2K

We recently bought a new microwave—our old gave up the ghost. It's shiny and has many buttons. It's happy not only cooking our food but telling us time and keeping time for us with a handy timer. But, unfortunately, while it wants to do all those new neat things for us, the amount of physical real estate for the button control panel is very limited. And so many buttons have to do double, triple, and even quadruple duty in order to provide us with access to all the functionality built into this device.

I once caught three doctors—these are people trained at prestigious medical schools, completed internships and residencies, and have been practicing for 20 years or so—arguing and pushing buttons on our microwave in futility, all to warm up a cup of coffee! I hate the damn microwave timer—it takes at least five button presses and a dial rotation to make it work! What were those product designers thinking? And don't let me get started on our new oven—it insists on knowing not only the time of day but the day and year as well and pretends to understand my recipes all in it's 2 by 3 inch window!

Mode errors result when a device has multiple modes of operation, making an appropriate action in one mode give an erroneous result in another. My microwave examples aside, this is the familiar remote control error—you press the play button for the VCR while the remote is in a TV mode and nothing works as expected. This is just another form of attention control error.

Unfortunately, as devices get smaller and smaller, designers rely on one button to do multiple actions, thus spawning numerous mode errors. And while they seem harmless (you can always just try switching modes), some people never get comfortable with the operation of some devices due to multiple functionality of controls. And in cases like car radio buttons which tend to do double duty as CD controls, drivers can get into accidents while they fiddle with mode controls.

Note that these kinds of errors are different from the ones when a person just doesn't get how to use their remote control—one is an attention error, the other in a background knowledge error. Understanding these types of errors makes it possible to try to design products which help users avoid them.

Perception Problems

If the doors of perception were cleansed,
everything would appear to man as it is, infinite.

—William Blake

Human perception consists of all the information that an individual receives from her senses: the visual, aural, tactile, and other perceptual information cues derived from the environment.

There is a push to introduce as many perceptual cues as technologically possible into an interface: gloves with force feedback, head sets with 3D stereo imaging, 3D spatial sound fields, and so on. But during problem solving, what a person notices is related to how she thinks things work—it's called **selective focus**.

At the computer, an individual needs to split her attention between the actions happening on the screen, the movements and clicks of the mouse, and the activity at the keyboard. A novice computer user will pay particular attention to only those actions that she thinks are causally related. For example, if a novice observer is watching a user open a file on a GUI computer, and the observer doesn't know the relationship between a mouse click and opening a file, then that part of the process is likely to be missed. The observer might generate an alternate but incorrect explanation as to how files get opened. They may conclude that simply putting the pointer over the icon object opens it.

My husband Christopher had a childhood interest in insects. He would often peer at insects under a microscope. But after obtaining a book on insect anatomy and learning the names for the various body parts, he noticed that he was observing more. It wasn't just that the observations were more precise, but rather it was almost as if learning the names of things enabled them to be seen at all.

One can think of this phenomenon in terms of the formulation: "The more you know, the more you notice; and the more you notice, the more you know."

Background Domain Knowledge and Errors

> *We are constantly misled by the ease with which our minds fall into the ruts of one or two experiences.*
>
> —Sir William Osler

Background Domain Knowledge Errors:

- **Misapplication of Problem Solving Strategies**
- **Causal Net Problems**
- **The Wrong Metaphor**
- **Erroneous Mental Model**

There is a saying that the most dangerous among us are those who know a little rather than those who know nothing. Armed with a little bit of knowledge, these individuals can destroy the world. There are plenty of problems when users suffer background knowledge errors. And like other errors, it helps to analyze their source.

One common source of errors comes from applying a "tried and true" solution to a problem that can't be solved that way—throwing water on a grease fire results in a deadly fire cloud, as water evaporates explosively sending burning oil droplets all over the kitchen. This is an example of **misapplication of problem solving strategies**. Firefighters hate that one.

Another common problem is misunderstanding the causation between two events—think rain dancing or blood letting. These are **causal net problems**.

Metaphors are common sources of errors. While a metaphor might help with initial understanding, it could also lead to mistakes when it is pushed too far—remember Senator Ted Steven's plumbing analogy for the Internet? Metaphors can also grow too old to be useful—how convenient is measuring the performance of the engine in your auto in units of horses?

And then there are errors caused by the mistaken understanding of how things work—**mental model** errors. Alchemy is a good example—lead won't turn into gold under laboratory conditions. Any time you visualize how something works, but it actually doesn't work that way, you are in a situation that could easily lead to errors.

Of course there are also situations when users just don't have enough background knowledge to use the product correctly. As always, careful product targeting is important to a product's success.

Misapplication of Problem Solving Strategies

From then on, when anything went wrong with a computer,
we said it had bugs in it.

—Rear Admiral Grace Murray Hopper, US Navy

Chapter 2: "Virtual Product Design Overview" discussed the "Interface as a Problem Solving Medium" approach to design. The gist is that we have to support users while they solve problems using our products and help them achieve their desired goals. We also need to understand possible reasons for failure.

One of the common mistakes users make is applying an incorrect or unproductive strategy to solve the desired problem. For example, in Math and Physics, certain classes of problems have very specific problem-solving strategies associated with them. The main difficulty in solving these problems is recognizing the class of the problem. Knowing what to do is most of the battle. Similarly, an expert computer user will be knowledgeable about what strategies work in a given situation—they know how to "troubleshoot" the system when things don't work the way they should. Novices, on the other hand, can be easily confused and lost when confronted with situations that are unfamiliar to them—they don't know how to solve the problem or even in what direction the solution lies. Misapplication of problem solving strategies is a major pitfall for inexperienced computer users.

A good product interface will "bring to mind" the right solution—it will point the user in the direction of success. A great interface would go one step further—the user would only wish to take one action, the right one!

Causal Net Problems

Causal net problems occur when people misinterpret a phenomenon and generate incorrect explanations of why things happen. For example, the notion that the Sun rotates around the Earth didn't go out of fashion until relatively recently. And if you ask a young child, you might still get: "The Earth is flat, and the sun is moving around and around the Earth, of course!"

We came across a causal net problem involving the use of a web browser by a novice. She kept trying to access a web site by typing its URL and pressing the LOAD button on the bar above. She was very frustrated because the only place she could ever go was the MSN Web site (her default home page). But the problem was that she never pressed the RETURN key after typing in the desired URL, and so the browser kept reloading the site that it was originally set on. Parenthetically, this same user also kept trying to contact companies by using their Web URL as an email address.

It's a lot easier to notice when you don't understand something than when you've misunderstood.

The Wrong Metaphor

The Linux philosophy is "Laugh in the face of danger." Oops.
Wrong One. "Do it yourself." Yes, that's it.

—Linus Torvalds

Here is an example of an awkward metaphor. I recently had surgery to remove my uterus due to a tumor growing in one of its walls. During the presurgical consultation, my doctor used a metaphor to try and explain why he should put me through such an ordeal: "Imagine an old worn out tire. Sure you can fix it, but at some point it's just not worth putting a patch on it—it's best to get rid of it." While my husband and I found this funny, I can envision situations where the use of an old tire metaphor for one's uterus wouldn't go over well.

At my post-op visit with the same doctor, he told me about the Neanderthal Diet—eat a little all day and lose weight. Again the image of a Neanderthal doesn't invoke the slender, graceful ideal that I was looking for in a diet—a name can make or break a product.

Erroneous Mental Models

Our beliefs may predispose us to misinterpret the facts, when ideally the facts should serve as the evidence upon which we base beliefs.

—Alan M. MacRobert and Ted Schultz

Do you know how your car works? If you're anything like I am, you'll have a general idea about pistons and mini explosions and such. And it needs water and oil, too. Oh, and I've seen under the hood—I definitely don't want to touch things there mostly because they are so greasy. What I'm saying is that I have a **mental model** of how the car works, but it's very sketchy and not very useful in an emergency.

A **mental model** is just what it sounds like—a mental representation of how something works. We have thousands of mental models of devices we use everyday: computers, calculators, cars, cameras, camcorders, cats, coffee makers, c-sections, coolers, spray canisters, rice cookers...I'm sure you're seeing a pattern here. You should, we humans are great at noticing patterns. Once you have a mental model of a device, for example a coffee maker, you can use it to make predictions about its behavior:

It's making happy gurgling sounds, it must be almost done making coffee.

The little red light is on, that must mean the power is on.

It's taking too long to brew. That could be an indication of a malfunction.

Mental models evolve based on observation. Unfortunately, as discussed previously, we are not very good at making accurate observations. We tend to pay attention to things we think are relevant and omit those that don't seem to be connected to the object of our interest. Thus we can easily decide that two events form a pattern even when they have nothing in common other then spacial or temporal proximity: Every time I wear that red dress, I get a parking ticket—there must be a relationship. Or more consequentially: Vaccinations delivered to 18-months-old kids have the potential to cause autism—we must keep children from being vaccinated.

Understanding the mental models that our users are likely to bring to our products is key for designers. We can help our users form more accurate mental models by creating diagrams or making action-consequence links more visible. Optimally, you want to maneuver the user into taking a single action—the only obvious and right thing to do with your product. Unfortunately, this doesn't happen often.

Have you ever gone to a friend's bathroom and weren't sure the door locked when you turned the little locking knob? The knob was happy to turn in either direction, opening the possibility of leaving you exposed at an inopportune moment. Did you ever look at a light switch and weren't sure which one would actually turn on the light you need? Did you ever have to provide technical support to your parents on how to switch the television setting from cable to DVD? Over the phone? All of these situations have something in common—a seemingly simple desire turns out to be a non-trivial, and sometimes difficult, action in practice.

Mental models can be very **simple**—that switch turns on that light. Or they can describe **complex** processes like human reproduction. A mental model can have excellent **predictive** power—a light switch physically located next to a light source controls that light source. Or it can be wildly off mark—having sex standing up is equivalent to using contraception. Just for fun, ask around for people's ideas on how a microwave oven works.

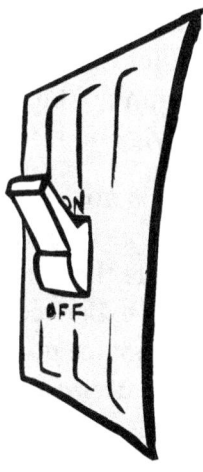

In Chapter 9 we discussed Background Knowledge. P-prims described little snippets of information that we gather in the course of our lives that describe and provide predictions of phenomena around us (e.g. that pot is steaming, it must be hot; but it could be dry ice). Like mental models, p-prims provide explanations and could be right or wrong and they might generate good predictions or bad. And like p-prims, mental models can be completely off base and still have pretty good predictive powers—keep asking those questions about how microwave ovens work.

Users have lots of mental models, but so do product designers. And we're not immune to generating wonderfully outlandish ones. But our mental models don't end with us, they get released into the world through the products we create. Take the following quote as an example.

Education is not filling a bucket, but lighting a fire.

—William Yeats

William Yeats makes a distinction between two mental models of the educational process. The difference between the "filling a bucket" approach to teaching and "lighting a fire" is not just metaphorical. Adopting one mental model over another has reverberating consequences from the kinds of books you buy to the activities you do in the classroom to the mental qualities you reward in your students. Mental models don't only affect how users interact with products. They also change how designers approach them.

Mismatched Goals & Motivation

Design is really an act of communication, which means having a deep understanding of the person with whom the designer is communicating.

—Donald Norman

A goal is an integral part of any problem solving situation. If you go to a supermarket with the intention of getting out of there as soon as possible, your approach to shopping will be different than if you go shopping with the goal of getting the most nutritious food available. Depending on the goal, an individual approaches the same task quite differently. Goals suggest the problem solving approach an individual is likely to take in a particular situation—check the food labels for nutrition facts or grab the needed items as fast as possible.

Goals can also change the way a person looks at the world and the objects in it. During dinner, a chair is just a chair. But when you need to change a light bulb, that same chair looks a lot like a stepping stool. Goals produce **opportunistic behavior** during problem solving. Think of building a Lego house or a sand castle—a wheel becomes a door handle and every twig and leaf can be used to good advantage in castle wall construction.

This goal driven opportunistic behavior is a large part of what people do on a computer. On a computer, the same goal can be achieved through many different paths. And the more agile the computer user, the more different ways she can think of to achieve the same goal. An expert computer user can use different applications to produce nearly identical output. A novice user may find that their output is dictated by the application they are using (e.g. if the user doesn't know how to change a font style, all his documents will have the same type style).

But when the goals of an actual user don't match the predicted goals of an intended audience for a product, there are problems. Then users often act in ways that are unanticipated by the designers, and the products fail. A current example might be the way that people are using browsers. Often users want to save the text of an article that has been coded into HTML. Copying and pasting text from a browser rarely provides the results that the users desire. At their worst, some browsers crash during these operations. Usually, they deliver a hopeless jumble of unusable text. (Yes, I know it's possible to delete all that junk, but who has the time?)

Kids' motivations for using a piece of educational computer software often doesn't match the optimistic goals that software developers have for those kids. It's entertainment versus education. When it comes to education, it is very common to have a big disconnect between the users and the suppliers.

Kids try to use computers to have fun. The designers are trying to create educational opportunities for the children. So consider an example: your job is to come up with a computer-based experience that teaches the alphabet. If you had to teach the alphabet to a non-English speaking adult, your goal and the goals of your prospective users match: everyone involved in the process shares the goal of having the user learn the alphabet. But if you are designing for a child, your task is doubled—not only do you have to come up with something that teaches the alphabet, but you also have to design a very entertaining way of doing it. Your goal and the goals of your intended users are inherently mismatched.

Here are some examples of children's behavior that comes directly from observing our own children at the computer. We use computers constantly—what chance do our kids have to escape their influence? When our youngest son was three years old, he had already been using the computer for two years. We have a large software library for kids, and our children went through a new software package every two weeks or so. We have one math game where the goal of the game is to add up the blocks as quickly as possible to reach the target number. The hope of the software developers of this game was to teach kids to associate numbers with actual quantities—in this case, with the number of blocks. Unfortunately, as soon as the kid reaches the right number of blocks, the round ends. Our son didn't like that. So instead of trying to get the right number of blocks, he played to get the wrong number, keeping the game going as long as possible—it was a lot more fun that way, and the animations were cuter. In their desire not to hurt the feelings of the "losers," the designers developed a better reward system for failure than for success.

In the example above, the **contextual goals** of the

developers and end users didn't match, and that led to the mismatch of the procedural goals. But while the ultimate goals—the contextual goals—for the developers and kids are sometimes different, the **procedural goals** could be made to be the same. What kids want to do on the computer should be closely matched to what the game developers want them to do.

It is always important to figure out **how** the product will be used and by **whom** and **why**. Many of the most important design decisions and solutions will come from understanding these three criteria.

Anxiety, Confusion, Frustration, Fear, and Fatigue

People everywhere enjoy believing things that they know are not true. It spares them the ordeal of thinking for themselves and taking responsibility for what they know.

—Brooks Atkinson

Never let a computer know you're in a hurry.

—Anonymous

When there is a lot of information that needs to be presented, there is a high risk of creating a lot of confusion and frustration. This a common problem among content-rich Web sites.

Most novices experience a lack of confidence when approaching an unfamiliar task. They fear breaking something unintentionally. This combination is bad for exploration. Users often fear that once they try something, they'll be committed to that action. These fears can be paralyzing because they impede users from experimenting and exploring—often necessary for a user to reach for tasks beyond their level of expertise. We'll be talking about some techniques to help overcome some of these problems.

Think back to how many times in your life you had to beg your computer to work for just a few more hours, until after the deadline. Remember all the promises you've made about "being good" to it later, whatever that meant. Extra pure electricity, perhaps?

Or have you ever heard things like:

"Computers simply don't like me!"

"Technology is not my thing."

"My mind turns blank when I read a computer manual."

Novices can't explain many of the things that happen while they use the computer. Things seem to happen for no reason. Often, they tend to make up explanations that cast the computer as a thinking entity, and often a diabolical one. These attitudes need to be understood and perhaps addressed when creating products for computer novices.

Memory Errors

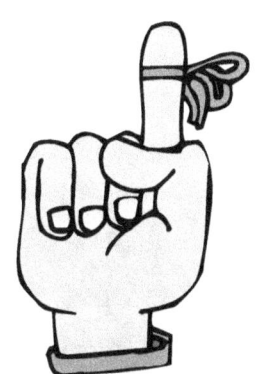

Eyewitnesses are notorious for their faulty memory of the events they saw. Ask five different witness of a car accident, and you will get five different accounts (at least in the details). And it's not just the case of perceptual blindness. Victims of violent crimes remember their assailants as taller by inches then they really are. And few victims try to remember little identifying details in the midsts of being victimized—in times of stress, the mind and body focuses on survival, not on reading tattoos. After the event, it's surprisingly easy to convince someone, still in shock, that a person responsible for their anguish is a big guy with a tattoo of a tiger across his face— that's just the kind of person who could do a thing like that.

Once an event is misremembered in a particular way, it's difficult to amend it. The more times you retrieve a bit of memory, the easier it is to remember it in the future. You form stronger long term memory bonds.

And except for a few memory wizards, most of us don't even register most of the information that we receive with our senses—that would put us into an informational overload. There is a wonderful experiment where an audience is shown a painting composed of identically sized, multicolored squares. Everyone is supposed to try to remember the image and are given a few minutes to do so. One of the squares is then outlined, and the audience is asked if the color of that square changed. The result is less than chance! We're just not very good at this. In fact, we do less well than chimpanzees. Recent studies in Primate Research Institute in Kyoto, Japan revealed that chimps have larger short term memories than humans, outperforming

us on certain tasks every time.

But what does poor memory imply for product design? We've already discussed problems with processing search engine results—all that jumping around and checking out search results for relevance overtaxes working memory and leads to lots of wasted time. There are other considerations as well.

Peek-a-boo Navigation

> *Computers make it easier to do a lot of things, but most of the things they make it easier to do don't need to be done.*
>
> —Andy Rooney

A common pitfall in Web design is what I call "peek-a-boo" navigation—it's when the functionality of a link is only revealed through a roll-over, if then. In order to navigate through these links, a user has to keep in mind the hidden information revealed through a roll-over and the position of that information for each option while trying to decide which link would be the appropriate one to take to reach the desired goal. And that goal also has to be firmly kept in working memory during this whole exercise. It's easy to see how this type of navigation can overwhelm working memory capacity.

Many corporate Web sites and Web design shops sport a peek-a-boo approach to design.

Environment and Errors

> *The World Wide Web has already provided us with a gigantic Information Marketplace, where individuals and organizations buy, sell, and freely exchange information and information services among one another. The press, radio, and television never got close; all they can do is spray the same information out from one source toward many destinations. Nor can the letter or the telephone approach the Web's power, because even though those media enable one-to-one exchanges, they are slow and devoid of the computer's ability to display, search, automate, and mediate.*
>
> —Michael L. Dertouzos

The amount of concentration one can bring to bear on a task depends, in part, on the number and quality of distractions. Environment affects working memory—anxiety and stress about a particular situation degrades an individual's performance. If a person is spending most of his working memory on worrying, then he has less working memory capacity to dedicate to the task itself. This is why some people are not good at taking tests—they might know the material, but the stress of the test situation diminishes their mental resources. And when a person has to perform in situation where failure is catastrophic—air traffic controllers, for example—it's the interface's job to make their work more cognitively manageable. In particular, the interface designer has to balance the speed of access to information—how deeply it is buried, how many clicks away it is—with the ease of recovering from an error.

Imagine another situation: a mother has a child who needs to go to the bathroom and is trying to use the informational kiosk to find the rest room. Given the child's urgency, the mother might not have the same cognitive capabilities as she would in her office environment. A kiosk in a zoo is not an ideal place or time to wrestle with complicated interface issues. You need super simple and clear navigation coupled with very spare screens. Think what a person needs to know, don't hide the ball under layers of information and screens or ads. If the kid has an accident while his mother is trying to figure out your interface, it's your fault!

Thoughtless Design

> *To arrive at the simple is difficult.*
>
> —Rashid Elisha

Sometimes I a get the feeling that designers of one device or another never actually use them themselves. My sister-in-law gave us a shiny, slick, and absurdly expensive tea pot. I love tea, it was a good present, but it's a lousy tea pot. After happily whistling to announce its readiness to serve hot water, it goes on to burn your fingers with hot steam. Every time! And not just my fingers, my husband tried to find ways to minimize the physical damage to our extremities, but to no avail. The thing was just awful! It was clear that the designer never tried to boil hot water using his creation. Most of his thought was clearly put into the look and feel—it looked and whistled great—but functionality was sadly neglected.

I've taught a class on product design on and off since the mid-1990s. And each time, on our show-and-tell day, the classroom fills with objects which design shortcomings students wish to enthusiastically expose in public. We've had alarm clocks that refused to be set; Web sites that brought down entire operating systems; rice cookers that almost burned down the house; door bells that would buzz randomly in the middle of the night; instructions for a toy that suffered greatly in translation; toasters that wouldn't release bread—actually, quite a few

of these were kitchen gadgets. People had elaborate workaround routines just to appease their devices.

It's clear that the total man-hours lost in productivity and anxiety over badly designed products far exceed the amount of time it would have taken the designers to do it right.

Additional Thoughts and Further Readings

In his book "The Psychology of Everyday Things," Donald Norman, a retired professor of cognitive science, provides wonderful examples of everyday objects which seem to have been designed to frustrate and humiliate their users. Norman goes on a tirade about doors that won't open and light switches that don't give clues to their operation. "Design is really an act of communication, which means having a deep understanding of the person with whom the designer is communicating," he writes. When that communication fails, errors occur.

For an interesting discussion about mental models and calculators, check out R. Young's paper, "Surrogates and Mappings: Two Kinds of Conceptual Models for Interactive Devices," which discusses the mental models of calculator users.

To learn more about the perceptual blindness experiment, visit the University of Illinois Visual Cognition Lab: http://viscog.beckman.uiuc.edu/djs_lab/

If you're interested in learning more about chimpanzees' short term memory feats, visit www.physorg.com/news115906589.html

Again, Mel Levine's book, "A Mind at a Time," is a great resource to learn more about the type of mistakes that children typically make and the corresponding cognitive causes.

18. Miscommunication

Wonder is the desire for knowledge.

—St. Thomas Aquinas

Miscommunication is quite a common occurrence—there are just so many ways things can go wrong. Let's start with the obvious—lack of common language. It's true, it's hard to communicate clearly when you don't speak the language, but millions of tourists do so everyday. Long discussions about the true meaning of life might be out, but you probably still can be understood enough to get a cup of coffee in a foreign land.

Simple and **concrete** ideas (refer back to our discussion on language) can be communicated with gestures and facial expressions. When my sister wanted a bit of honey in her tea, she was able to make buzzing noises and flap her hands next to her body in a convincing bee imitation, and a nice Portuguese waiter was happy to oblige her wish. While language-free, this exchange did require two willing participants, one expressive and inventive enough to be able to communicate without using words, and one patient and accommodating enough to guess at the meaning. But if my sister was a bit too shy to mimic a bee in public, or the waiter too busy, this communication would not have taken place. **Temperament** makes a difference.

If ideas we want to discuss are **abstract**, requiring **higher** order language skills, gestures and pointing just won't do the trick. Consider a situation where you have to discuss medical treatment for cancer. You and the doctor could well share a common language, but the words she's using must still be unfamiliar. **Background knowledge** in medicine might be required to understand the nuances of the treatment options being proposed. It's very easy to lose track of understanding in such situation. What's more, sometimes, it's not obvious that you've misunderstood

Product Adaptation Categories:

- **Physiological**

- **Perceptual**

- **Cognitive**

- **Content Focus**

until after the conversation is over and you no longer can ask for clarification. Not only didn't you understand what was being said, but the doctor might not even be aware of the miscommunication. This is a double failure, and clearly more dangerous.

Now consider the same situation, except the doctor is telling you that you are the one with cancer. Take the time to soak this in—you might die; it might hurt; it might take a long time; your family will be devastated. Your heart rate should be going up now and you should be sweaty and uncomfortable. You're anxious. And you're still in the middle of the conversation with your doctor about possible treatment options. The likelihood that you will be able to comprehend what is being said and be able to make rational decisions based on the information is rather low. Miscommunication is common when participants are placed under **emotional stress**.

Let's think of a happier occasion. You're visiting New York City and language is not an issue—your control of the English language is impeccable. You find a nice stranger and ask for directions to MoMA. The stranger is happy to oblige, gives long and detailed instructions, all while the city workers rip up the street with a jackhammer. Make that two jackhammers. You politely walk away with only a vague idea of where to go. **Environmental conditions** can hamper communication.

All that jackhammering gave you a splitting headache and probably permanently damaged your hearing. MoMA's nice tour guide is enthusiastically discussing Picasso's influence on Georges Braque's Cubism, and vice versa, while you're only catching every other word. Your **perception** has been compromised by loud sounds and handicapped your understanding.

Language Barriers

Language provides access: access to information, access to community, access to social status, access to employment. Language is embedded in culture and inseparable from it. We talk about street language and scientific discourse—both might be based on the English language, but the vocabulary, the turn of phrase, and even the sound of the words are quite different. It's easy to speak the same language and still be unable to communicate. To join a community, to become a member of a social group, you have to learn its language.

Product interaction is a form of communication: it's a dialogue between the product designer and her audience. To be understood, product designers have to speak the language of their audience. They have to become part of that community of users, at least temporarily.

Communication Barriers:

- **Lack of common language**
- **Colloquial variations**
- **Insufficient vocabulary**
- **Incomplete subject matter knowledge**
- **Cultural differences in tone, emotional style**
- **Divergent communication styles**

As with other characteristics, it helps to break down communication failures into specific language barriers. The first, and most obvious, is the dominant language used by the product's audience (e.g. French, English, Russian, Japanese, etc.). If you're not a native speaker, it's important that your translator is. You'll avoid a lot of embarrassment that way.

The history of product design is full of such sad examples of names working in one language but not another. There was a popular Japanese soft drink with a name that to English speakers sounds like "Cow Piss." Or the "Chevrolet Nova," a perfectly acceptable name in English, but which means "Chevrolet doesn't go" in Spanish ("no va" translates into "no go").

1. prohibition against 3 years old below of child usage;

2. play attention. you of finger,hair,clothes...etc.don't touch and car wheel,inorder to prevent quilt harm;

3. car while driving not want to by hand grasp it;

4. don't let the remote control close to any fire with car original:(such as electric stove, stove beside or mightiness of sunlight bottom)

5. not want the place in danger to play;(such as street, steep slope...etc.)

6. don't let the wet water of car, and not want under the rainy day is open-air usage;

7. mot want on the sand ground to play;

8. forbid the child to tear open the remote control with the car;

9. if the car dash to plecesed, and should pass by the person check or profession personnel maintain the rear can continue to use.

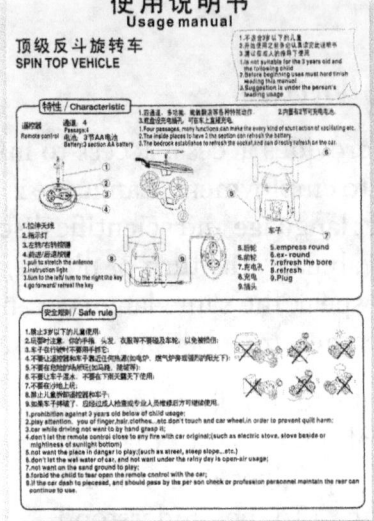

The example of toy instructions shown here clearly suffered from translation. The large text, numbered one through nine, is left as is, with its lack of spaces and liberal sprinkling of punctuation.

While television and movies have smoothed out a lot of regional language variation, different parts of the world speaking the same language manage to be very inventive with their vocabulary and turn of phrase—colloquial variations can quickly sour communication. Simple expressions used locally to mean something completely different can result in socially awkward moments (e.g. consider the phrase "cutting the cheese").

When you can't find the right words to describe what you mean, the outcome is more than just failure of communication. Words help us remember—if you can't describe the details, chances are you won't be able to remember them later. When used with a product, if a person is forced to refer to "that thingy" too many times, interaction becomes strained—the dialogue between the product designer and the user breaks down.

The weather might be a safe topic of conversation unless you happen to be talking with a meteorologist. Some communication requires higher level language skill and a very technical vocabulary. Limited knowledge of word meanings can not only cause a local misunderstanding, but can easily lead to a poor business outcome. We've all been in situations where a person we're talking to is too polite to say "no" to our request. That person might even use positive, affirmative words, all the while meaning quite the opposite. The result can be total breakdown in understanding. Cultures in which manners might prevent individuals from saying directly what they mean can be classified as practicing an **Indirect Communication Style**. At the risk of oversimplifying, peoples of Middle East and Asia tend to use this communication style. On the other hand, Americans, including Central and South America, are more direct—they don't let manners interfere with expressing negative opinions or decisions. Americans tend to have a **Direct Communication Style**.

Another dimension by which to analyze communication is the amount of emotion infused into speech. North American and Asian populations tend to be more restrained and tend to refrain from grandiose displays of emotions. We practice a **Restrained Emotional Style** of communication. Latin Americans and peoples of the Middle East lean more toward an **Expressive Emotional Style** of communication.

By plotting the two dimensions on a plane, we get a **Communication Style Field**. A product designed for an audience in a particular quadrant of this field needs to adapt the desired communication style of its users.

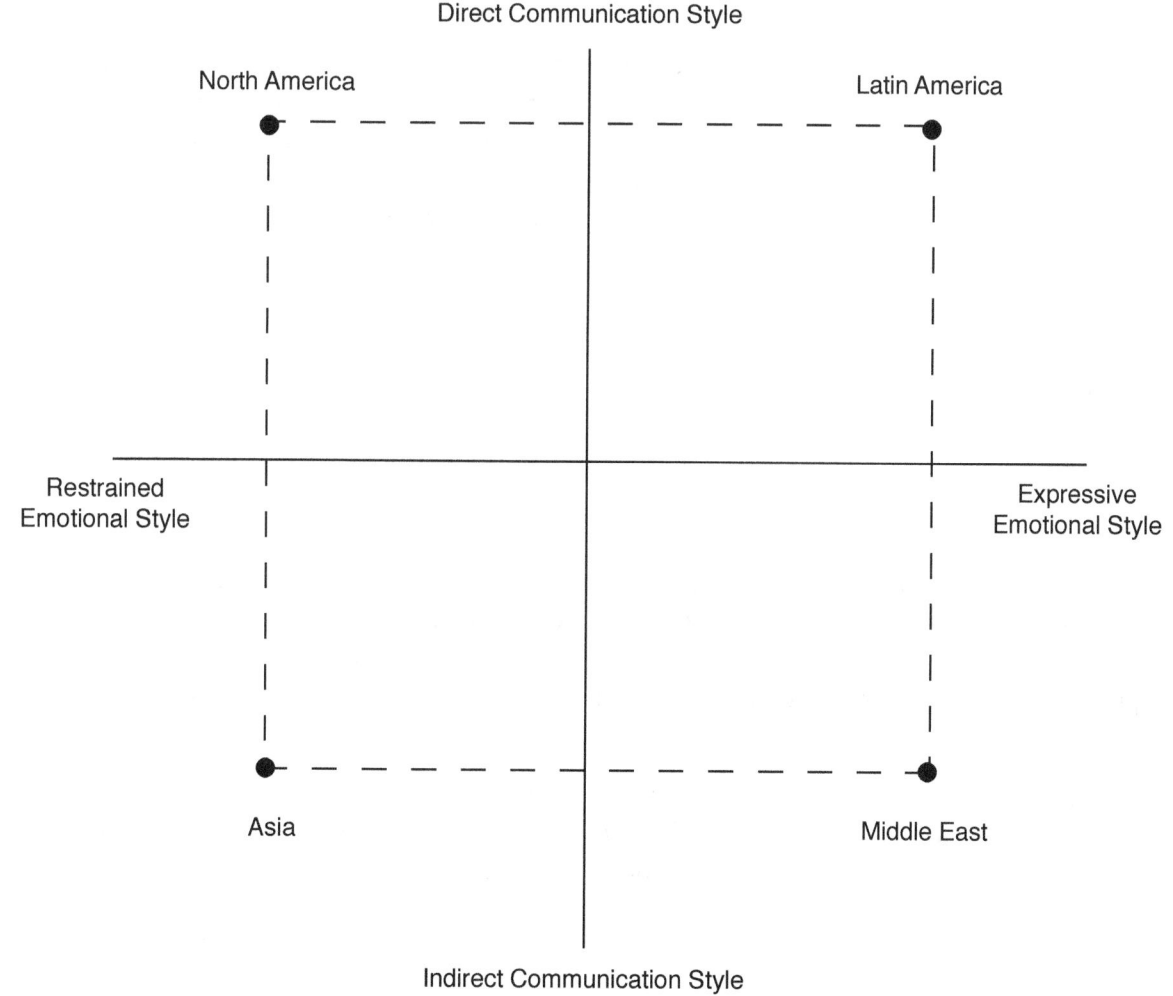

Cultural Errors

He who proves things by experience increases his knowledge;
he who believes blindly increases his errors.

—Chinese Proverb

They say in France it's not polite to keep your hands and elbows under the table—not only do you want to show off your jewelry, but you want everyone to see that you are keeping your hands to yourself and aren't fondling other diners' companions under the table. At my mother's dinner table, it is the height of bad manners to lean your elbows on the edge of the table. Manners, like so many other design criteria, are highly culture specific.

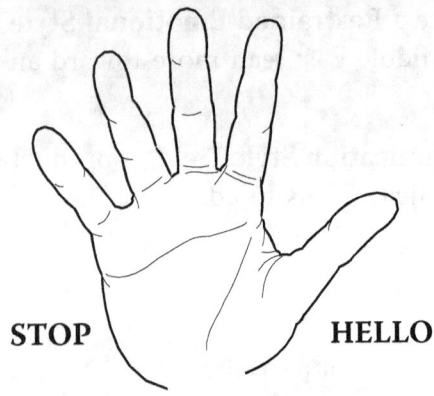

STOP   **HELLO**

Offending someone due to lack of culture-specific manners may sour first impressions with a product. But more serious consequences can result due to misinterpretation of intentions. It was heart-breaking to read about Iraqi civilian deaths that resulted from the cultural misunderstanding of simple gesture: a hand raised by a soldier to signal "stop" was inferred to mean "hello" as per local customs.

As a product designer, it's your job to make sure that your products are used as envisioned wherever they are marketed. There are many instances of medical equipment being used outside of the original culture of its target audience, for example. A simple label, a set of visuals, written directions, and other methods to clearly designate appropriate interactions can save lives.

Social Snafus

The Internet is the Viagra of big business.

—Jack Welch, Chairman and CEO, General Electric

The difference between cultural errors and social snafus is clearly a bit arbitrary, but helpful. Cultural errors focus on the differences between large groups, differentiated by political borders, language, and history. Social errors result from local variations and customs among people speaking the same language, living in the same country, and sharing the same history.

Most obvious opportunities for social snafus involve the dinner table. As a kid in Russia, I would get into serious trouble for speaking at the dinner table. Now in America, kids are expected to talk during dinner. The use and misuse of cutlery is a standard way of judging someone's social graces. Can you pick out the desert spoon?

And for an extra bonus, what is this? How is it used? What clues does it have as to its function? A hint: it opens and closes.

A well-designed product can minimize the social discomfort of its users by suggesting the appropriate way to engage with itself: a handle, a switch, a bit of instruction can go a long way.

Mismatched Scripts

The mind is like the stomach. It is not how much
you put into it that counts, but how much it digests.

—Albert Jay Nock

San Francisco's Exploratorium has a persistent problem with its visitors: a lot of the people who go there are first time visitors to this type of museum. These visitors have a long standing museum script: "Don't touch the exhibits." But at the Exploratorium, the whole idea is that everything is hands-on. If you go there and stand and watch people interacting with exhibits, you'll often hear parents scolding their children for "inappropriate" touching and handling of objects. It's the Exploratorium's job to change the museum scripts of its new visitors and show them a new way of acting in a museum. (See the upcoming section "Permission Giving and Modeling" in the next chapter for one possible solution to the mismatched scripts problem.)

Another example of a script that needs to change is "don't give out credit card numbers over the Internet." There is still a lot of resistance to typing in the credit card numbers into secure browsers among certain demographics. Some of these same people don't have a problem giving their card numbers to a merchant over the telephone. There is a lot of money being spent by companies doing e-commerce in an attempt to change this particular script.

Myth Propagation

Things are not always what they seem.

—Phaedrus

Very few actions are performed in isolation. Many individuals prefer to spend some time observing others prior to engaging with an object. These "observations" are no longer necessarily direct—the Internet, newspapers, televisions, and other media sources all contribute to the creation of expectations prior to an actual experience.

Some formal background knowledge may be acquired under dubious circumstances and end up just being wrong. And since what an individual notices in any given situation is highly dependent on his social and domain background knowledge, the resulting causal net generated prior to an actual experience can be faulty. This "fault" does not have to be unique to a particular individual, but rather can be constructed and propagated by a group of users. Hence **Myth Propagation**.

I wrote a paper on this phenomenon based on ethnographic observations of visitors interacting with a Tornado Exhibit at the Exploratorium. Prior to taking their turn at this exhibit, these visitors would observe the active users of Tornado Exhibit walking around the perimeter of the rotating column of gas. The myth developed that it was the action of these walkers which caused the rotation and consolidation of the vortex column. And this belief was passed on to the next group awaiting their turn at this exhibit. It wasn't until the exhibit stood empty that the myth chain was broken.

Mythology, in the context of computer use, is when one person's misunderstanding of a computer-based event or an action propagates through a community of users. Mythology is learned and passed on from individual to individual through various types of interpersonal communications—email, conversation, letters, phone messages, etc. There are just so many ways insure that your misunderstanding is shared far and wide.

Mythology is of course not limited to computer use and can cover any subject area. Urban legends are prime examples of mythology—inaccurate background knowledge distributed through a population.

Negative mythology about a product can cause anxiety, fear, and confusion among people who are not familiar with it. Some mythology can be quite harmful. For example, the notion that computer use can be bad for child's brain development is not true, but it scares a lot of parents into denying their children a very good educational tool. In fact, any new technology tends to generate fear and misunderstanding among some portion of the general population. When books first came out, many considered reading detrimental to one's health. Negative mythology often propagates from person to person until enough people become aware enough about the truth of the subject so as not to believe it.

Additional Thoughts and Further Readings

The Internet is a great way to get on the net.

—Senator Bob Dole

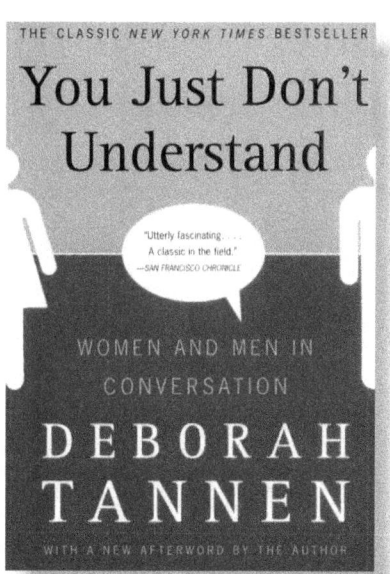

My paper on the Exploratorium was titled "The Relationship Between Changes in Perceptual Focus and Understanding" and was presented at the American Education Research Association (AERA) conference in 1994. It is available on AERA's Web site.

For an interesting foray into conversational differences between the sexes, consider reading Deborah Tannen's "You Just Don't Understand: Women and Men in Conversation." Deborah Tannen's other book, "You're Wearing THAT?: Understanding Mothers and Daughters in Conversation," captures the relationship between mothers and daughters through a linguistic prism.

For a bit of fun and practice, please analyze the following error:

What do you imagine happened in ordering this cake?

19. Design Recommendations

Designing an object to be simple and clear takes at least twice as long as the usual way. It requires concentration at the outset on how a clear and simple system would work, followed by the steps required to make it come out that way—steps which are often much harder and more complex than the ordinary ones. It also requires relentless pursuit of simplicity even when obstacles appear which would seem to stand in the way of that simplicity.

—T. H. Nelson (1977), "The Home Computer Revolution."

Simple and Clear Interfaces are Hard to Design

Up to 95% of all baby car seats are installed improperly. Whose fault is it? The parents certainly want to keep their children safe, right? But have you ever read the instructions on the side of a baby car seat?

If we always remembered everything we ever experienced, and never misunderstood instructions, and didn't get anxious or overwhelmed by circumstances, the life of designers would be easier. There would still be problems due to poor design, but at least users wouldn't be contributing to failure. But people do forget, get tired, and their attention slips—people make mistakes. And designers have to understand those mistakes and their origins and design to minimize them.

From time to time, we all find ourselves in situations where we just don't get it; things slip away, and our goals feel just out of reach. Faced with these situations, some of us feel bad, some angry, some frustrated. Some blame the product, while others think it is all our own fault. How we feel about failure depends on our personality. Our approach to overcoming failure has a lot to do with the strategies that worked for us in the past.

There are three distinct approaches a designer can take when developing a product for a specific audience.

One, try to always create materials in the cognitive "style" of the prospective users. That is if you know you are dealing with graphic artists, chances are that most will have a strong preference for visual information, and so the product can be made to cater to that perceptional style.

Two, if you know that the product doesn't match the cognitive style of the prospective users, surround the product with auxiliary content that would support those users in reaching their goals despite their limitations.

Third, if you know your users will have to operate in situations that are difficult for them, create materials that teach them to overcome those difficulties.

You can think of this in terms of learning. As a student, you can be taught in a cognitive style that makes learning easier, faster, and more fun for you. As an alternative to individualizing instruction in the above manner, the regular school curriculum can be augmented with supplementary materials that would help you understand it better. And finally, you can be taught strategies to overcome your cognitive limitations and learn to deal with the educational world as it is.

All three approaches are valid and useful under appropriate circumstances.

At the beginning of every design process, the following questions need to be considered:

- Who is the primary audience? What are their goals for using the product? Why do they form a group? What do they have in common? What is a list of cognitive characteristics that defines this group?

- Who is a secondary audience? What are their goals for using the product? And so forth. Don't stop here. Identify all possible users—there is no need to stop at two groups.

- What are the circumstances under which the product is to be used? How much time do users have to reach their goals? What are the environmental pressures exerted on the users while they are trying to reach their goals?

- Which combination of individualization, support structures, and learning materials is appropriate for the design of the product? This question needs to be asked three times: once for the conceptual design stage, once for the interaction design stage, and once for the interface design stage.

- What is the desired learning curve and the target proficiency level?

In some very real ways, the beginning of a design cycle is all about creating taxonomies: who are all the users and what are their strengths and limitations, and what are the circumstances of use. Once the users and the environment have been sorted into groups, the designer of the product has a better fix on the rest of the design process.

Analysis of Failure

Remember: you are the only person who thinks in your mind!
You are the power and authority in your world.

— Louise Hay

I once had considerable trouble getting the point of purchase terminal at a gas station pump to accept my credit card. Even with a diagram, I was having problems orienting my card correctly. Given the opening, there were four possible ways to insert the card. The diagram wasn't good enough to limit my choices and the other interface elements took a long time to notify me that my card was not accepted. Was there something wrong with my card? Or was I doing something wrong? Or was the system broken?

How can we examine my actions given the conceptual, interaction, interface design model? Conceptual design deals with **what** the product does: I knew what was expected of me. I was required to pay for gas. No failure there.

Interaction design deals **how** the product does what it does: I was supposed to stick my cash card into the slot. The card fit just fine. Again, so far so good. Or is it?

Interface design deals with the **look and feel**—that's the part that should be informing me of how well I'm doing with the device: While I knew that my card was rejected, I couldn't figure out why. There's the failure!

It would have been wonderful if the card could be inserted into the device one way only. A little groove or a key-like notch would have done the trick, and thus my problem could have been solved at the interaction design level. But this solution would require changes to cash and credit card design—unrealistic.

So we're back at looking for a design solution at the interface design level, right? But what about imbedding a little sensor that figures out the orientation of my card inside the reader? Then the machine can tell me to flip the card—this would be on interaction and interface design solution, the feedback being part of the interface design in this case. Sounds okay as long as the cost of production doesn't make this uneconomical. This is a technological solution.

Even better, is to make the device accept my cash card in any orientation—just put in the card and voilà! It's all working perfectly as long as I remember my password correctly. From a usability standpoint, this would be best. From an economic point of view, it depends of how much each extra sensor adds to a cost of producing the card-reading device. If failures like mine are common, then the cost-benefit analysis might swing in favor of extra sensors. If my problems are unique, then of course spending more per device doesn't make sense. That's where the need for user research comes in.

A cheaper solution might be to improve the ancillary materials which instructs the user on how to orient the card. This is an interface design solution. Perhaps a better design could substantially solve the problem. Maybe the screen, which can display video, could show how someone inserts the card in the correct orientation.

User Personas

> *All generalizations are dangerous, even this one.*
>
> —Alexandre Dumas

Product design begins by creating taxonomies:

- Who are the main users of the product?
- What are they trying to accomplish? What are their goals?
- What are their cognitive strengths and limitations?
- Under what environmental and technical conditions are they using the product?

These questions are not specific to a particular industry. The product could be a museum exhibit, a car, a Web site, or a toy. Regardless of what it is, a product designer has to determine the answers to these questions.

Usually, multiple product audiences are identified during the initial conceptual design meetings. For example, if a product is a Web site, these users can be:

- Casual visitors
- Current and potential community members
- Current and potential customers/buyers of products and services
- The press
- Regulators, reviewers, and inspectors (relevant to some businesses)
- Business partners—potential and actual
- Employees—current and potential

There could also be ex-members and ex-employees that still have some sort of continued relationship with company, either forced or by choice. As you can see, it's easy to start making categories of visitors once you get down to it. And each of these groups of visitors have their own needs, goals, environmental conditions, and cognitive abilities.

Note that there can be an overlap between user groups. In fact, users are constantly migrating from one taxonomic category into another: from casual visitor into a community member; from a member of the press to a company's employee. Fortunately, the needs and goals of various visitors from various user groups tend to overlap: contact information and directions are needed by many users; images created for casual visitors can satisfy the community members and the press; information about employee benefits can help users make up their minds about becoming members; and so on. But there are also significant differences in tone, language, and vocabulary appropriate for each audience.

The problem with taxonomies is that it's hard to understand and relate to individuals belonging to a particular group. It's easy to start thinking of the "Press," perhaps, as a hostile entity instead of as an individual trying to do a job. **User personas** solve this problem by helping product designers and stakeholders visualize their users.

So this is a two-stage process. The first stage lists the general characteristics of each user group. The second stage creates a set of fictional well-defined personalities that represent each group. These personalities are used by the design team to visualize the needs of the audience. It's easier to visualize the Zone of Proximal Development and Flow conditions for a particular, well-defined albeit fictional individual than for group.

Here's an example of a user group definition and a user persona for a fictional travel company described at the beginning of this book:

Individual Visitors: Members, Prospects, Casual Surfers

The first set of audience members are Individual Visitors. There are three types of visitors that fall into this class: members, potential members, and casual surfers (i.e. individuals that will never become members). While it seems obvious why it's important to satisfy the first two, the casual surfer also needs to be wowed. Casual surfers can spread the word, create buzz, help start viral marketing, and generate press and recognition. There will be many more casual surfers than there will ever be Africa Safari Tours members or even potential members. Thus they count, too.

Audience	Goals	Cognitive/Environmental Background
Casual Surfers	1. Look at materials available for free. 2. Research travel destinations without joining Africa Safari Tours. Most probably the majority of these visitors are coming from search engines. 3. Make a decision to investigate further—move into "Potential Member" category of visitors.	1. All ages, both genders, all background levels. 2. Visitors referred from search engines that land on a secondary page of Africa Safari Tours' Web site need a quick way to identify and orient the context in which the content is presented. 3. All configuration of hardware and software. 4. Can't make assumptions about connection speeds.
Potential Members	1. Get background information about the company and its products including the company's history, leadership, vision, technology, and so forth. 2. Determine the value of services and do a cost/benefit analysis for their family. 3. Research customer reviews and satisfaction. 4. Get contact information. 5. Request additional information. 6. Find membership benefits. 7. Make a decision about becoming a member.	1-2. are the same as above. 3. Visitors can either be referred to the site, stumble upon it, or find it through advertising—different attitudes towards the company will be based on the initial encounter situation. 4. Users access the Web site both from home and work. 5. Users access the Web site from the "field"—accessing information from hotels, Internet cafés, on the road, and at camps—a vast differences in computer resources and connection speeds. 6. A vast difference in experience with adventure travel: both sophisticated and novice travellers and everything in-between. 7. Vast differences in experience doing self-reservations for a complicated trip. 8. Need to accommodate differences in background knowledge about various destinations: cultural, language, historic, etc. This defines the background knowledge relating to particular locations.

Audience	Goals	Cognitive/Environmental Background
Current Members	*Before the trip:* 1. Browse multiple vacation possibilities in a fun and kid-accessible way. Have a fun time visualizing themselves in far destinations. 2. Research information for the next family vacation. Get inspired for the next family trip. 3. Choose a travel destination—help family come to consensus on a particular destination. 4. Get the latest information about a particular vacation. 5. Get travel advice for a specific destination given the unique family constraints (e.g. small kids, physical disabilities, time constraints, season constraints, political constraints, financial constraints, psychological constraints, etc.). 6. Get tools to plan a particular vacation: schedules and time tables, reference numbers and contact information, medical and bureaucratic checklists for the whole family, packing suggestions. 7. Seek information and help to solve an idiosyncratic problem. 8. Buy materialists/kits put together by the company for a particular destination. 9. Get recommendations or reviews from other Africa Safari Tours' members about a particular destination. 10. Get contact information.	1., 4., 6., 7., and 8. the same as above. 9. Need to support parents of little kids. 10. Need to support decisions made over a long period of time and by multiple members of the family: parents and kids might make/have different criteria about what's important in a vacation; different family members might browse at different times and for different durations and need a different amount of support to make their choice. 11. The site needs to be fun for parents and kids of different ages—while using Africa Safari Tours site, members should not be bored or feel that they are working too hard.
	During the trip: 1. Get the latest information about a particular destination. 2. Resolve a problem with Africa Safari Tours' travel partners. 3. Look up contacts, locations, or other type of information for a particular location. 4. Get customer support. 5. Get contact information.	1. and 5. are the same as above. 12. Stress of travel: time zone changes; cultural shock, physical discomforts, language barriers, etc. 13. Stress due to idiosyncratic problems: lost travel itineraries, miscommunication with travel partners, medical emergencies, etc.
	After the trip: 1. Compare and share experiences with other members. 2. Offer suggestions or criticisms. 3. Create a scrapbook of the family vacation. 4. Start thinking about the next trip. 5. Get contact information. 6. Post a review.	1. and 4. are the same as above. 14. There is a strong desire among people to share experience among their communities—Africa Safari Tours can be a starting place to facilitate the post-trip experience. 15. People tend to reconstruct their experiences, Africa Safari Tours can help create positive memories of family trips by providing souvenirs of the trip and other supporting materials.

Below is fictional personification of one of these families designed to make their needs and goals concrete to the creators of Africa Safari Tours' services. It's beneficial to the product to discuss concrete scenarios of use—it keeps both the design and the designers appropriately grounded and focused on the needs and desires of their audience.

Ella and George Katz—Current Members, Before the Trip

Mr. and Dr. Katz and their eleven-year-old son Jonathan joined Africa Safari Tours about six months ago. While they are already seasoned travellers, they are looking to improve their experiences and take more control over their vacations. The Katz Family are sophisticated consumers—they know a lot, they have travel expertise, they will be quick to form a judgment. They don't have time to waste on razzle-dazzle. They want to make sure that they get the value from their membership. Ella and George are relying on Africa Safari Tours' reputation and judgment to reduce some of the risk associated with planning a difficult vacation without the assistance of a travel agent.

Jonathan is also old enough to check out Africa Safari Tours' options for himself. He wants to make sure that he would have a good time on his vacation (not always the case in his experience). He would enjoy finding out how other kids really felt about the trip his parents are planning for him.

Each member of the Katz family will surf the site on their own in addition to exploring it as a group.

The Katz family walking on a beach.

This example shows how to move from the general characteristics of a particular user group to the specific needs and goals of "real" people. Instead of discussing "Current Members prior to taking a trip with Africa Safari Tours," the product designers can now talk about the Katz family: "Ella and George would never book a trip without discussing it with Jonathan first." This is a whole different conversation. And it's much better for the end product.

The Home Page of a Web Site

The beginning is the most important part of the work.

—Plato

Home pages are tricky. Small companies are trying to look big; big companies are trying to look accessible or hip; graphic shops are trying to display their technological chops. A company's ego is always on the line with their home page. Some companies seem to care only about getting the color of their logo exactly right (they say it's essential for branding). Some are just interested in having something move—which may explain all those letters-folding-into-envelopes-stuffed-into-mailbox animated GIFs that seemed to self propagate through the Internet in the late '90s like a virus.

The first thought should always be the needs of the audience. Perhaps a Web design shop can convince itself that its audience of prospective purchasers of Web design services really needs to see all that graphical gimmickry so that they know just how capable they are. We've tried to convince ourselves that that's true, because the whiz bang stuff is often a lot of fun to design and execute.

But beyond the branding needs of a company or a Web experience, the keys to a home page are twofold—it should make clear what content the site has to offer and it should offer an easy way to get to that content. Content is king on the Web. It's what the audience is there for in the first place.

Sometimes an audience is so truly divergent that their needs won't be easily met by one home page. Our company was hired to consult on one project where the users really were of three separate and distinct types. We came up with a solution where the home screen allowed the visitors to identify themselves with a particular group of users and thus self select which section of the Web site was appropriate to them. You have to be very clear and help the users figure out to which group they belong. The U.S. Government's Web home page is of this type: www.usa.gov.

Work in Progress

A Web site is always a work in progress. Business' needs revise. Technology evolves. Users' goals change. Nothing is set in stone.

There's a continuous cost-benefit analysis: Is the cost of adding a new feature covered by improved usability? Will the Web site redesign result in increased business? If the new home page is better than the current one but not all that you were hoping for, go with the new design—if it's better, if it's cost-effective, there's no need to wait for perfection. Tomorrow the world might change again, and you will need to reconceptualise your Web needs one more time. Continuous incremental improvement is more achievable goal then perfection.

Web work is not set in stone—a site is always a work in progress.

Graphics, Visual Memory, and Understanding

Seek simplicity and distrust it.

— Alfred North Whitehead

Keep backgrounds very simple. And just because something is big on the page, doesn't mean that it is visible. Simple shapes are more memorable. This relates to the memory limitations as discussed previously. An interesting example to consider is the design of traffic signs. These images have to be read and interpreted by a driver in a fraction of a second, as she glances at them to make navigational or other traffic-related decision. And traffic signs strive to be culturally independent—since countries honor each others' driving licences, an individual navigating unfamiliar streets in an unfamiliar state relies on the visual information in these signs to understand not only navigational directions but traffic laws which might differ from those to which they are accustomed. From personal experience driving in unfamiliar foreign countries, some traffic sign are more successful in their endeavor to communicate information than others.

Layout for a Computer Screen

> *[There] are many more wrong roads than right ones.*
>
> —Jeffrey P. Cohn

The goal of good layout is to draw attention to the most important elements of the page: navigation, logo, contact information, new items, or a product release notice. For graphical recommendations, it's good to consider the wisdom of newspapers, magazines, and comic books designers.

The Web creates additional challenges as compared to paper layouts:

- the size of the page is not well specified,
- the look of the page may change when viewed with different browsers and on different computers,
- the color choice is only an approximation of what the user will see,
- "below the fold" means "off the screen"—design needs to encourage scrolling or, alternatively, stay within one screen height,
- design needs to accommodate technical limitations of creating Web pages—a Web page is more like a software program than a brochure,
- users have control over text size and some font choices,
- graphical elements can move (e.g. video and animation can be incorporated into the page layout),
- there are multiple ways of arriving and leaving the page.

First the layout versus the computer screen: To scroll or not to scroll? In most cases, scroll. On the Web, users don't mind scrolling—it's easier than clicking back and forth trying to find what you're looking for. Remember, recognition beats cold recall in ease of use. Moving from page to page requires keeping information in working memory, remembering items that were located on previously visited pages, and having a rudimentary grasp of the site's information architecture. This is a lot to ask of a casual visitor or even sometimes of a determined one. Providing a lot of information on a single page allows the visitors to use recognition rather then relying on recall. Users tend to scan pages for trigger words upon which they can click and which they think get them closer to their goal.

Clearly, making your Web page wider than an average computer resolution screen, or stuffing it with giant graphical elements so it takes too long to load is not in your best interest—if your visitors don't get to see it, what's the point of putting it up? So there are technical and practical limitations to computer screen layout.

As mammals, we're programmed by millions of years of evolution to notice movement above all other perceptual cues. If it moves, we have to pay attention to it. That's why all those flashing ads are so annoying—we have to notice them, we can't help it. Creating elements on Web pages that move on purpose but just for the sake of having something move is no different from the annoying ads. Movement has to aid comprehension. It needs to add rather than take away from perception of information. If you have something move on your Web page, make sure you know why: What are your goals?

Captions provide random access to information featured in this book. They encourage surface reading. While that's good at a book store, while the buyer is looking for the right book to buy, it's not to be encouraged in a textbook. Thus there are very few captions. But there are plenty of subheads—these form the informational outline of this book's content.

From research done on newspaper readers and direct mail pieces: most people tend to read captions first, then subheads, and only then do they get around to reading the headlines. So to increase the information transfer rates, make sure that the key talking points are featured in the captions. You might want to repeat those same points in the subheads, and even create "asides" that show them again as a bulleted list.

There are many graphical ways to solve the below the fold layout problem. The main idea of all of them is to insure that the reader doesn't think he reached the end at the visible boundary of the screen. Diagrams, columns of text, photographs can all span the fold and insure that there's no illusion of the end of content at the bottom of the screen. Horizontal rules that span the width of the screen act as "stoppers" and tend to discourage users from scrolling past them.

Content can encourage scrolling.

Chances are that if you picked up a newspaper, it's because you both chose to take a look at what's inside and that you know which paper you're holding. While surfing though, you might end up on a page of some site by following a few links. The thread of information that leads you to a page might be perfectly reasonable, but the end result could be surprising—hyperlinks are like that. On the Web, it's easy to slide into content sideways and avoid the home page all together. This puts design pressure on secondary pages of the Web site to include identifying information in the layout. Visitors have to quickly and easily orient themselves in relation to content: What is the point of view? Who is the author? Am I being sold something? How do I navigate this site?

Information and Content Architecture for the Web

The nice thing about standards is that there are so many to choose from.

—Andrew S. Tannenbaum

The job of an information wizard is to specify and organise the content that will go up on the Web site in view of its audience' needs and expectations. That's a tough job.

Content architecture identifies the bits of content and details how they need to be constructed—each bit of content needs to be identified, saved in a usable format, edited, approved, and assigned a particular and unique location on the site. This isn't just the obvious big blocks of articles and copy, but also all the captions, headlines, alt tags, legal information, and other content which slips between the major chunks of content. Each bit of content serves a particular purpose for a particular audience. If a paragraph of text is meant to satisfy a particular goal, it needs to have the right tone, length, and content to do so.

Information architecture organizes the content materials destined for the Web site into a structure that best fits the conceptual design. If content architect specifies the content that needs to be created, information architect sculpts it into a useful and usable form via navigation, searchability, and organization.

To this end, information architect designs:

Information Architecture Tools:

- **Navigation**
- **Searchability**
- **Organization**

- Labelling and tagging system of each bit of content
- Organizational system for all of the content
- Navigational structure for the whole site
- Table of contents and lists of graphics
- Site maps
- Visual organizers
- Content guides
- Contextual hyperlinks
- Content search system
- Frequently asked questions, terms, and information
- Superheads, headlines, subheads that group content together
- Glossaries and indices
- Bibliographies
- Metadata and meta tags
- Sequential lists and step by step instructions
- Check lists
- External links

If you have problems finding the information you need, some of the blame rests on the shoulders of the information architect.

Information Arrangement for the Web

Users can't be expected to grok your Web site information architecture completely. Environmental conditions, background knowledge, attention controls, reasoning styles all work against you. Adapting industry standards helps—most users have learned the categories in their previous interactions with other Web sites. Understanding the expectations of the types of users you are likely to get is very important—doctors group information very differently from patients, for example.

A Web site is a hypertext document—a set of multi-linked pages. For usability purposes, hypertext information should either be **deep and narrow** or **shallow and broad**.

If the information is arranged to be **deep**, then the user is only presented with a few choices at each depth level, minimizing errors due to inattention and limited working memory capacity. In such an arrangement, if a user makes a mistake, it is easy to back up and return to a known position.

Slideshows are a classic examples of deep (and one-wide) navigation: the user just needs to click "next," "next," "next." Some online learning systems use this structure for linear content presentation, but usually for their "deep" material—once the user has arrived at a particular content, the presentation turns narrow.

In the **broad** and **shallow** arrangement, most of the choices are visible on the screen at all times—the user doesn't need to rely on recall but rather uses recognition. A user is only a few clicks away from most of the content on the site. Ideally, each major area of content is a single click from the home page and most of the content is no more than two clicks away. The two-click navigation system supports fragmented attention span and limited background knowledge that some visitors bring to Web site exploration.

Broad and shallow navigation schemata and deep and narrow schemata both keep disorientation due to environmental interruptions to a minimum. Incomplete understanding of the materials can also lead to miscomprehension of information. Again by limiting the site to either one approach or the other, visitors won't have to use much of their cognitive resources on navigation.

Unfortunately, it is very easy to slip from sticking to these information architecture arrangements. Over time, different departments or product lines expand, and the site starts to grow organically. **Organic growth** is neither broad and shallow nor deep and narrow—it's like a big, thick bush. If you hear: "Our visitors don't use our navigation to get around—they just use the search engine," then it usually means that your information architecture got tangled, and the only way your visitors can find anything is by using the search function. It's time to hire an information architect again!

Standard Content Organization Schemes

When there is a lot of information and it is arranged using a broad and shallow architecture, the breadth of the site becomes large. Consider the display of search results in Google, for example. While auto-generated, the search results page presents information in a very broad and one-deep structure—each result is a link with short tag description of its content, and the list of results can run hundreds of pages long. A good search algorithm tries its best to put the most useful links up at the top of the search results, but that still leaves a lot of choices for the user to sift through. It would be great to be able to sort the search results based on some additional criteria like date, author, place of origin, etc.

Individual Web sites also have problems with breadth. To mitigate the number of selections a particular user will have to examine to find the desired information, the Web home page should group all content into general areas. There are standard groupings:

- Alphabetical
- Sequential—information encoded by time
- Geographical

- Geopolitical

- Subject Matter—based on large groupings like politics, mathematics, biology, literature, etc.

And there are idiosyncratic groupings:

- Categorical—based on categories defined by either the company whose content this is or by its users (e.g. check out Google's image tagging game at http://images.google.com/imagelabeler/); these are smaller groupings like Civil War, Set Theory, plant morphology, Orson Scott Card, etc.

- Thematic

As with most of the ideas explored in this book, thematic and categorical organizational schemata are culture-depended and tend to have a lot of individual variations—what's forms an obvious set to one group may not to another. Usability studies help with identifying pattern variations based on cultural and micro-cultural differences.

Organization of information deals with creating sets of information. For example: fun activities include games of chance, which subsume poker, which encompass Texas Hold'em, which contain the rules of that game. Because sets can overlap, organization can get muddied, and users can get lost in a navigational tangle. Moreover, because information now resides in several places (duplicated), it becomes difficult to maintain and update. This is a common problem. Usually, avoiding it requires one clear-thinking gatekeeper who understands the organizational scheme and insists on it being followed.

Metaphors

Grade school is the snooze button on the clock-radio of life.

—John Rogers

A metaphor's job is to make an unfamiliar situation seem akin to a familiar one—metaphors are a form of scaffolding. The metaphor chosen by the designer influences the expectations, goals, and perceptions of its users.

In the case of the Mac, a desktop metaphor was chosen. Instead of the abstract command language of DOS or UNIX (i.e. COPY C:\WINDOWS\TEMP.TXT A:\ ), the desktop metaphor invited the user to open a folder and drag an icon in the shape of a piece of paper onto a picture of a floppy disk. Hard drives are file cabinets, directories are file folders, and files are pieces of paper. There is even a trash can.

But a desktop metaphor has its share of problems. There are places where the Mac desktop metaphor, before system X, was unduly stretched. For example, to eject a floppy diskette, users were asked to place its image in the trash. Many balked, and even experienced users report feeling a slight unease as they visualize their work being trashed.

Like the Mac (and many say because of it), Microsoft Windows is also an example of a WIMP interface. And there are places where the Windows metaphor breaks down as well. For example, on Windows 95, many users have difficulty with the concept that when they want to shut down the machine they have hit a button labeled START.

Both Mac and Windows interfaces seem to share the following basic problems:

- Novices have difficulty locating saved files.

- Novices have problems understanding and distinguishing between different device locations (floppies, hard disks, servers, etc.).

- Novices have difficulties with windows: not seeing all the icons in a window, scrolling within a window, etc.

- Novices frequently complain about all the files that suddenly appear on their desktop.

- Novices frequently lose files under open objects.

- Novices have difficulty knowing what applications are currently loaded into memory. In the Windows interface, this often results in users launching programs twice or thrice. Users also have difficulty switching between programs, or even knowing that they can.

- Novices have trouble with the menu bar changing when applications change.

- The "Search Space" of the Mac and Windows interfaces is large, and novice users experience cognitive difficulties in understanding where and how to search for their application or file.

With both Windows and the Mac adopting a desktop metaphor, it's easy to assume that it is the only WIMP interface possible. But the desktop metaphor isn't necessarily the best one for a variety of users. For example, young children are unfamiliar with office tasks, so that metaphor does nothing to enhance their understanding.

There are a number of possible metaphors that could be applied to the interface design of an operating system. Here are some examples:

- The Boxer Operating System was designed by Dr. A. diSessa et al. at the University of California at Berkeley. This system also uses a WIMP structure but is based on the metaphor of boxes within boxes—boxes contain various objects that the user creates. Boxer is evocative of a construction set and raises expectations along those lines.

- The Cyberspace Model or Geographic Metaphor is based on a 3D landscape that the user navigates (e.g. the fantasy interfaces depicted in the movies "Jurassic Park" or "Disclosure").

- Factory Metaphor—where applications are thought of as processing and molding the initial idea of the user into its final realization. The factory-based metaphor focuses on workflow and process.

- "Time Strings," David Gelernter's alternative operating system—where everything is continually carried backwards in a chronological flow like a river and keywords pull out sub-time strings, all organized chronologically (e.g. look for files that were created last week). In some ways, the "stacks" feature of Mac OS X 10.5 "Leopard" partially implements this idea

- And then there are Internet Browsers which see all files and folders as pages.

The use of metaphor in interface design is not limited to operating system software, of course. One example is Bryce 3D, an application whose function is to allow the user to create images of three dimensional landscapes. While Bryce 3D uses windows, icons, menus, and pointers, its underlying metaphor has nothing to do with a desktop. The Bryce 3D interface is complex and powerful, and it allows a person who is just starting to learn this program to create very realistic and beautiful landforms. And since the creation of landforms wasn't something that was previously done anywhere, including the top of a desk, there wasn't any reason to adopt the desktop metaphor. In fact, adopting that metaphor would have made the program more cumbersome to use.

Painter is a program which harnesses amazing algorithms to generate bitmap images which resemble natural painting tools—brushes wet with oil paint, for example. Painter uses a real-world metaphor for its functions, choosing an artist's studio. Brushes are housed in "drawers" which are even decorated to look like they're made of wood. Painter requires that users prepare a canvas prior to beginning work. Prior to making a mark, users must choose a brush tool in one section, adjust its size in a second section, and adjust its transparency in a third. But Painter failed to appreciate that many of its users were going to have previous computer experience with other graphic design applications. Failing to capitalize on that previous experience meant that professional users had to learn a complex computer interface from scratch.

Consider desktop publishing programs as another example. These programs are designed for users who may previously have performed page layout functions manually. These prospective users are familiar with X-acto knives, waxers, columns of type, leading, kerning, cropping tools, points and picas and agates, rolls of plastic lines, and grid sheets. These users might not be familiar with digital color management techniques, resolution issues, digital font management, threaded paragraph frames, and a host of other computer-related issues. A good desktop publishing interface builds on what graphic layout artists already know. For example, page layout programs frequently use icons derived from the tools used by graphic artists in manual layout. A good desktop publishing program has a foundation rooted in what users already understand how to do manually, and also creates a supportive environment in which they can reach beyond what they know how to do on the computer. See the section entitled "Zones of Proximal Development and Maximum Benefit" in Chapter 9: "Background Knowledge."

When computer applications are built to emulate or replace manual tasks, there is a danger in excluding those users who are just entering the profession and are new to manual as well as computer-based tasks. Care must be taken to create interfaces that are not weighed down by obsolete metaphors. And all metaphors can be stretched too far until they become meaningless to most users.

For example, although typeset columns used to come on long sheets of paper and were cut to size with an X-acto knife to fit into a layout, it would be stretching the metaphor in a computer program to require that the user designate the end of a column of type by employing the X-acto knife tool. If this was done, a new user would have to understand arcane manual layout techniques before they would find the appropriate tool in a computer interface.

Another example from Photoshop, whose interface I greatly admire by the way, is the "Unsharp Mask" command. This is a sharpening tool. It is called Unsharp Mask because of some arcane procedures in printing photographs in a darkroom that most professionals now only dimly recall. But unless the user specifically knows what the Unsharp Mask command is for, it's probably the last one that would be chosen by a user looking for a command to sharpen his image.

KidPix, which is otherwise a good art program for kids, also has an annoying interface element—an image of a moving truck was used as a tool to pick up pieces of a drawing and transfer them somewhere else on the page. Many kids can't figure it out because it is so far removed from what they do when they draw pictures on paper. KidPix stretched the metaphor too far and lost its audience.

Metaphors can be used for navigational purposes as well. The most common distinction in Web-based navigational metaphors is between link-based navigational systems and geographical ones. In a geographic system, the spatial organization of items on the screen parallels some real world environment, providing navigational clues to its users. In a link-based system, navigational items are either provided as text links or buttons. While the spatial organization of the link-based system may provide clues on navigating the web site (i.e. important links are at the top), there is no attempt to bring to the mind a real world situation.

Planet Oasis, for example, adopted a geographical metaphor. Users began with a bird's eye view of a city and can click on city blocks. They can zoom in closer and closer to their desired target—from a city block to a particular building.

Children's CD-ROM software titles often adopt a geographical metaphor. Kids are familiar with their local geography—they have first-hand experience with getting from their house to the playground, to school, to a friend's house, and to nearby stores. Kids also associate different places with the different activities that they can do there. So an interface based on a geographic space often makes much more sense to kids then a desktop metaphor. And this metaphor relieves some of the necessity of providing text labels that can't be read by pre-reading children.

Culture and Metaphor

We constantly make analogies and create metaphors: "It's like a river—time just flows along in one direction." "Life is a one way road." "Time is an arrow." "Time is relative." "Time flows." "The circle of life." Time is a difficult concept, and we have many metaphors to "try to grab hold of it"—this too is another metaphor for capturing an idea—can an idea be captured? Not only do different cultures develop different metaphors for the same idea, but as cultures evolve through time, these metaphors change. Finding these cultural variations is part of the background research that product developers need to do prior to launching the design effort.

Metaphor Use Recommendations

The following recommendations for metaphor use are adapted from a 1982 article, "Metaphor and the Cognitive Representation of Computing Systems," by Carroll and Thomas:

1: Find appropriate metaphors for each user group. What works for a five year old might not work for a sixty five year old.

2: Given a choice between two metaphors, choose the one which most closely resembles the system you're trying to develop. The more aspects of the system that can be covered by a single metaphor, the better.

3: Ensure that the emotional tone of the metaphor is appropriate to the desired emotional attitude of the user (think of the gynecologist's metaphor to describe my uterus: "old worn out tire").

4: When it is necessary to use more than one metaphor, choose metaphors drawn from one domain (i.e. close enough), but not so close that these metaphors represent exclusive alternatives (i.e. not too close).

5: Consider the consequences of your choice of metaphor for users and designers.

6: Show the users the limitations of the chosen metaphor.

7: Metaphors are useful learning tools, relating one unfamiliar system to another. But at some point in the mastery of the product, users might find that the metaphor no longer serves a useful purpose. The designer should then let it go.

8: Provide the user with exciting metaphors for routine or boring work and eventually present the user with a variety of scenarios which present different views and require different actions but whose underlying structure is identical.

Creation of User Expectations

One is tempted to define man as a rational animal who always loses his temper when he is called upon to act in accordance with the dictates of reason.

—Oscar Wilde

A well-designed interface should be easily and entirely understood by its users. What users don't understand, they won't use.

A well-designed interface allows the user to easily find the functions of the program that the user wants to utilize. An intuitive interface allows the user to correctly guess both the existence of and the location of functions in the program.

For example, an artist would expect a graphics program to have a "rotate" function, and she should be able to easily find the location of this function. A more prosaic example is a door—most door users expect the door to open and close and should be able to easily locate that "functionality" in the specific door example they are encountering.

A cool feature can sometimes have a poor interface—or at least a poorly explained one.

A few years ago, some public bathrooms started installing faucets without knobs. Water flowed automatically when the user simply placed their hands under the faucet. People are used to knobs on faucets. The lack of knobs created a lot of confusion, and only through pure accident, and then through word-of-mouth (or permission giving), was the interface to knobless faucets finally understood. A temporary note of instruction or even a graphic above the faucet would have solved this problem. In this situation, while conveying the operation of this interface was poorly done, the audience was captive—people really wanted to wash their hands after using the public bathroom. When it comes to cryptic computer interfaces, designs can't assume that the incentive to understand them is as high.

The problems of the knobless faucets with no instructions is particularly amusing considering the endless instructions given on hot air hand dryers: "Press button and rub hands gently three to four inches under the flow of hot air." At one bathroom, a piece of graffiti scratched underneath read: "Then wipe on pants." The real difference, I suspect, is that the hand dryer provided more obvious real estate to affix instructions.

One of our clients developed a pay-per-use Internet device that allowed anyone with twenty five cents to access the web. These devices used a touchscreen and were intended to be placed in bars and other public places. An alpha version of the device was put in a busy bar and the development team sat a few tables away to secretly observe people's reaction to their creation. One businessman approached the device and placed a coin in. Then he stood back and waited. And waited. But nothing happened. After a few minutes, he walked away. The development team was dying inside: "Didn't this guy get it? Why is he not touching the screen?" The answer was that he didn't expect this to be a touchscreen, and nothing on the screen led him to believe it was. Our recommendations included adding a video "attract" mode which showed happy people touching the screen to interact with the device.

Shaping of Goals

> *These are not the droids you're looking for.*
>
> —Obi Wan Kenobi

People are lured to a Web site to find information or to solve a problem. But once they've entered a site, it's up to the site's developers and designers to guide their audience and shape their goals and expectations. E-commerce sites hope to entice their visitors to buy something. News sites hope to keep their visitor's attention long enough to show them as many page views and advertisements as they can. Community sites hope to gain a new member and, perhaps, a contributor to their content.

It happens quite frequently that people have their goals changed and manipulated by the media—they go to the store to buy one thing, but they leave with a bunch of sale items they'd never considered buying before entering the store. It's the same on the Web.

Permission Giving and Modeling

> *Any science or technology which is sufficiently advanced is indistinguishable from magic.*
>
> —Arthur C. Clarke

> *Any technology that is distinguishable from magic is not sufficiently advanced.*
>
> —Gregory Benford

A person involved in a novel activity will be more likely to try something new if directly encouraged to do so.

Permission giving is a surprising and important component of human-computer interaction. It comes out of my research on how visitors to the Exploratorium—a museum of science and human perception in San Francisco—interacted with hands-on science exhibits. On any given day, the Exploratorium has many first time visitors—people who have never seen or interacted with its exhibits. On crowded days, each exhibit has a large group of people observing the lucky few who actually get to put their hands on the controls. Those at the controls become the de facto demonstrators of that exhibit. The more imaginative those demonstrators were, the more interesting were the ways that exhibit was used. The better the demonstration, the more the observers learned.

But in a hands-on environment, many people are timid. Perhaps they are afraid of breaking something unintentionally. Or perhaps they don't want look stupid. And so they constrain their actions. For example, the Exploratorium features a Tornado Exhibit. It consists of a large cylindrical booth with fog produced beneath the floor and a fan inside the ceiling. The fog is sucked up to the top of the exhibit forming a mini tornado. On crowded days, there are usually a few individuals that actually climb inside this exhibit and walk around the tornado. And after they are done playing with this exhibit and leave, others are always eager to jump inside and walk around the tornado, too. Their behavior continues as a chain of permission giving to each future set of visitors. But on days when there are only a few visitors, most people just stand and watch the tornado go around without jumping in and touching it. These people are afraid of breaking something or acting inappropriately. Most people are not used to touching exhibits in a museum; their museum script says "no touching." European visitors are especially vulnerable to "hands off" scripts, having more experience with fine art exhibitions rather than exploratory environments of hands-on museums. People need permission to touch or, in this case, climb inside the exhibit. If a few people do go inside the tornado exhibit and walk around on an uncrowded day, there is no one waiting to emulate their example. The chain of permission is broken. Each future group of visitors must decide if getting inside the exhibit is appropriate, and most decide it is not.

Similar behavior can be observed when a group of people use a piece of software or a Web site.

A good interface should take into account the aspects of a product that may be too unique to grasp independently by its users. These product features therefore require a way to give user explicit permission to interact with them in the intended manner. There are numerous ways of achieving this. One is to create a demo mode for a product that shows a particular way of using it. This would generate a "script" in the mind of the user about how to interact with the product and would shape their expectations. Most video arcade games have an attract mode, which acts both as an advertisement and a subtle tutorial.

Cultural Differences

> *[W]e are shaped by the sort of society in which we live, and we would not be the same person if we had grown up elsewhere.*
>
> —Robert Sapolsky

The number four, according to Western numerology, is associated with the four corners of the earth, home, stability. To the Chinese, it means death. How we see the world around us, what we pay attention to, what we remember, how we interact with people around us is in large part determined by the culture we were born into and by the culture we live in. I was born in Russia. My family immigrated to America when I was thirteen. To a certain degree, decades later, there's still culture shock. My parents will always believe in every conspiracy theory they encounter—the government (doesn't matter which one) is always out to get them. I will still flunk the "Russian Spy Test," as my husband calls it: Tiddlywinks? Meathead? In today's society, it's the little things, easily taken for granted by designers, that can completely subvert a product's functionally.

The problem with cultural differences is that they are difficult to analyze: Is that person not touching the exhibit because he believes it's the "wrong" thing to do? Or does he simply have greasy hands from eating popcorn earlier? Ethnographic research can answer some of the usability differences, but it's expensive and time consuming. So it's always back to cost-benefit analysis: How valuable is the data? How much is it worth? What's the downside of product failure?

Fortunately, there are companies that now specialize in obtaining ethnographic data for micro-populations. Marketing and political campaign executives routinely use information like "white wine drinkers are more likely yo vote Democratic" while "people who eat rutabagas are prone to vote Republican." Some institutions gather their own data.

Museums tend to collect information on their members. There's also data on the number of visitors per day and which special exhibits tend to attract large crowds. If your goal is to develop a new exhibit, looking at some of this data and then actually spending some time at the museum can inform your design. If you do this for a living, you won't be able to stop paying attention to how people interact with exhibits even during your off hours, and this will further enrich your development process.

Parks, zoos, shopping malls, and supermarkets are not much different from museums. Visitors form certain expectations and goals prior to going to these locations, and they walk away happy if those expectations are met. Each of these public places has a few **focal points**—locations that draw the most number of individuals. Bathrooms, checkouts, ticket counters, concession stands, the Mona Lisa—they're all focal points. Designers need to plan for these focal points and develop systems that ease congestion, relieve stress, and increase the enjoyability of the interaction.

Cultural differences become apparent at points of ambiguity, authority, and unanticipated stress—how people deal with long lines, for example. We were in London during the terrorist attacks on the bus and the underground in June 2004. While there was a lot of stress, locals and tourists behaved very well. There was a basic trust in authority, and people were very accommodating to emergency workers and others out in the street. The British Museum remained opened for most of the day, even though one of the blasts was just two blocks away. Would a different culture handle this situation differently?

While we relate to others in times of crisis, we still tend to view the situation from our own cultural bias. Why would that Russian woman tolerate abuse from her local policeman? He works for her, doesn't he? How could America keep so many of its citizens in jail? How can Kenyans turn a blind eye to corruption by government employees? The answer tends to be the same: "Because that's how it is." As a product designer, you just have to understand what that "it" is and accommodate it accordingly.

A user and product designer communicate via product interaction. It's a conversation. And this conversation can be complicated by cultural differences.

Group Projects

Collaboration, cooperation, and competition can be explored from the point of view of an individual working on a particular project. In **collaborations**, everyone knows each other and each other's capabilities, groups tend to be small, and projects limited in scope.

In **cooperations**, some group members know each other and some don't. Cooperative groups can be large with members having limited knowledge of what others are doing, and group members can come and go during the project's tenure. But individuals in both cooperative and collaborative projects share goals for the overall project and contribute their work towards achieving those goals. As group projects, newspapers, magazines, journals are all cooperative—they feature authors and attribute work to individuals even if overall these publications are group projects. Authors could compete among themselves for accolades set by the norms of their communities.

Competitions are different in this respect. Competitors don't share any of the workload among themselves. They might not know the individuals that are competing against them or what they are working on. There is no shared information or work. The individual competitor's goal is to beat out the others. Typical classroom interaction tend to be competitive, with students ranked by their achievement levels.

Project Type	Group Size	Scope	Duration
Collaborative	a small group	limited in scope by the size of the group	limited in duration: based on the intersection of time availabilities of all project group members—all tasks have to be worked on jointly by the group
Cooperative	a large group	unlimited in scope: from large to small	unlimited in duration: based on the combined total (sum) of available time contributed by all the members of the group—tasks are distributed among the group (members can come and go)
Competitive	a single individual or a small group competing with another individual or small group	limited by individual's productivity	limited in duration: based on the total amount of time an individual can devote to a project

This chart summarizes the classification of group projects into collaborations, cooperations, and competitions using the main variables: group size, scope of work, and project duration.

Group dynamics play an important role in social product design. Group dynamics specify whether the situation is collaborative, competitive, or cooperative and note the important conditions for these environments: the size of a group, the duration of the project, the scope of work, the individual time availability, the distribution of expertise among the group, the social status of group members, the rate of communication among group members, and the distribution of work among the group's members.

Online Learning and The Company Therapist Project

Education costs money, but so does ignorance.

—Sir Claude Moser

Online learning takes on a wide variety of forms, and the number of learning opportunities available is continuously increasing. Museums are putting worksheets for students and tutorials for their visitors on the web. Government agencies are transferring guidelines and tests online. Corporate organizations are uploading their training materials onto the Internet. Educational institutions are making their courses available to their online students. Such a large collection of learning opportunities developed for different ages, in various formats, and with different purposes makes it possible, in theory, to customize learning not only to fit individuals' goals but also their learning preferences. But what makes one computer-based instruction successful while others fail?

It is critical to match the needs and goals of students with those of online instruction—this is goal alignment. Adult students already seek out educational opportunities that best fit them.

This is the front screen of The Company Therapist, as available at www.TheTherapist.com

This is an example of graphical support provided to a writer by the project: schizophrenic woman's doodle.

Young students pick high schools and colleges that can satisfy their economic, social, and academic needs. This is also true of students looking for an education on the Internet. As an example of a successful online learning class, I would like to introduce you to The Company Therapist project. It was a creative writing course designed for a small number of students who self-selected themselves to fit the format of its instruction. The Company Therapist was designed and produced by Pipsqueak Productions, LLC. An archived version of the site is still available for viewing at www.TheTherapist.com.

All projects have goals and design constraints. Broadly, The Company Therapist project had three instructional design goals: to design and produce an educational structure which was capable of supporting a long-term community of students; to use this structure to teach creative writing; and to extend the project beyond the student participants and open it up to a larger audience of Internet browsers.

Design constraints force limits on the project, but they also provide inspiration and force the development of creative solutions. Since we were the producers of The Company Therapist project, we had a very limited budget and time availability with which to work—the project couldn't be too expensive for us to produce, it couldn't take too long to develop or take more then a certain number of hours per week to run. We decided that it would be free to the participating students and that their time burden couldn't be more than a few hours per week. Although the students would contribute writing, they wouldn't be responsible for the HTML production, interaction design, or the graphical elements on the site. Illustrations, editing, commentary, formatting, and audience were extra bonuses not available in the regular creative writing classes.

We also understood that creative writing students we would be working with would have the following characteristics as a group:

- they would be of varying skill levels,
- they would have widely different time availability,
- they would be geographically, culturally, and socially diverse,
- they wouldn't know each other,
- and they would only communicate with each other via email.

We wanted to provide these individuals with a membership in a long-term writing community; to give each amateur writer exposure to a professional environment (deadline-driven, well-defined standard of quality and format, editorial support, etc.); to help participating students develop their writing skills using a series of instructional support structures; and to get recognition for writers' work by exposing them to a large audience. The solution that fitted all of these design criteria was a *hyperdrama* written cooperatively by its audience.

Hyperdrama is a new literary form possible with the advent of computers and particularly the Internet. *Hyper* refers to the non-linear nature of this format. The stories are developed through an interlinked series of essays. The reader is usually free to skip and jump around, between, and within story lines. While such hyper writing can be presented in a book format, the Internet adds an ease of navigation from place to place with just a click.

From a product design perspective, hyperdrama's unique literary structure is well-suited for group writing. Each writer can contribute a piece to a greater whole and in the process gain experience and improve writing skills. All student writers cooperate to create an overall literary landscape and all feel pride in the total project. Yet individual accomplishments are easily separated from the whole. It is easy to spot good writers and to find prolific contributors to the overall project. So while all enjoy the benefits of multiple awards and positive press coverage—The Company Therapist project earned numerous awards and was recognized for its accomplishments by its reader, writers, and the Internet community at large, winning Entertainment Site of The Year in 1997—each writer can still point to his or her individual work product. This structure solved a lot of inter-personal frictions often present in other collaborative writing projects.

Turning the product of a creative writing course into an online entertainment opened the project to a much larger community of readers, writers, and reviewers than would ever be possible in a bricks and mortar classroom, thus turning student work into an authentic writing experience.

Turning readers into writers had another important side benefit: it increased readership. As readers decided to contribute their own content to the site, they told all their friends and family about their new hobby. And these personal communities which enveloped each author were not only motivated to visit the site, but also to visit it regularly. Thus, each new writer not only added to the overall content, but expanded the audience for all the writers of The Company Therapist project. For a site that depended on readership for survival, such increase in visitors was critical.

In deciding on the main story line for The Company Therapist project, we realized that we needed to create a world which was easily understood by writers contributing from around the world (most writers for The Company Therapist didn't live in the San Francisco Bay Area). Thus, we chose present time and modern society. From the Internet user demographics, we knew that the vast majority of users in 1996 were over 18 years of age. These users were early adoptors of the Internet and had computer access at home or at school.

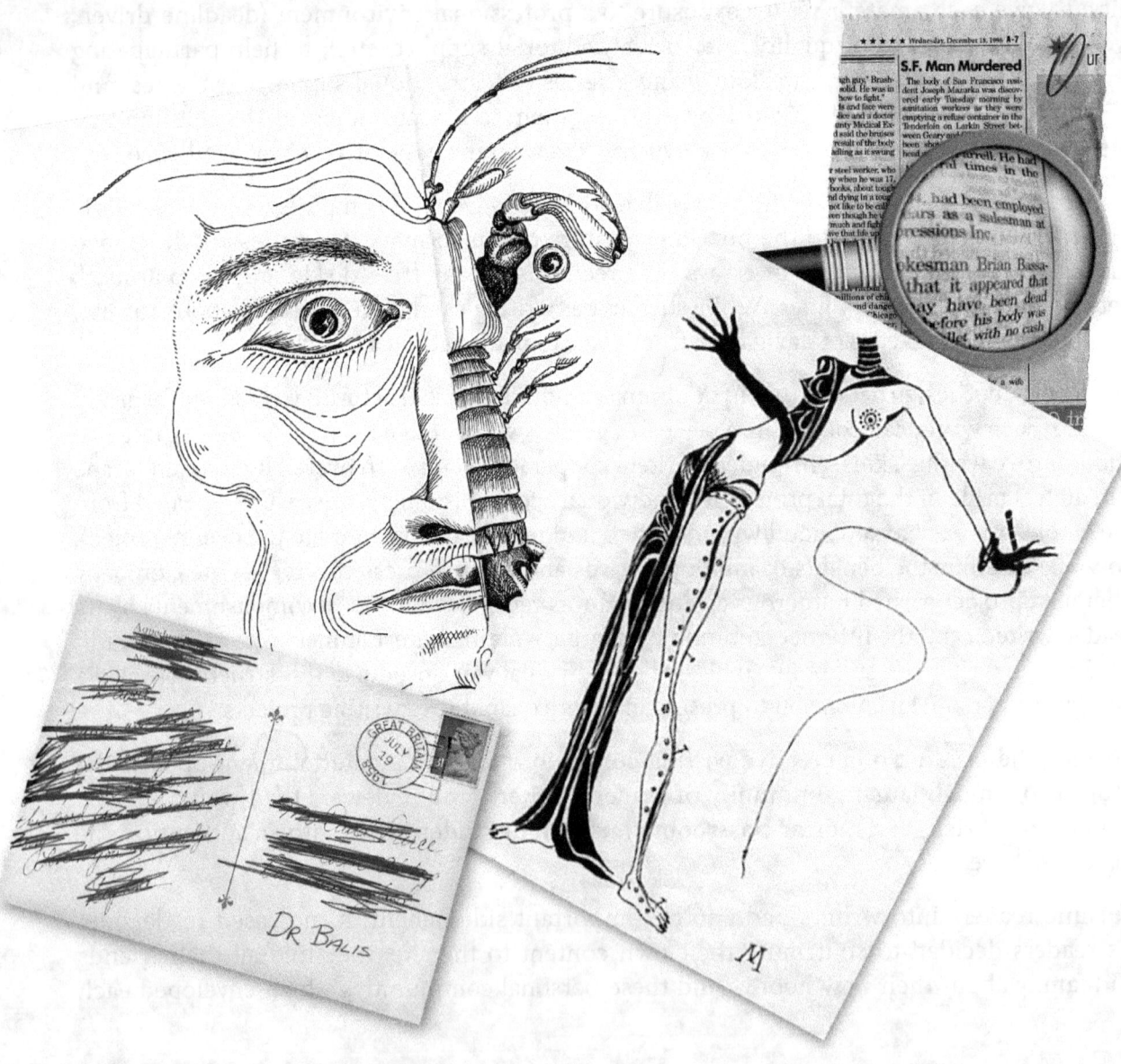

The notion of a writing in a form of a psychiatrist's dialogue sessions developed to give the most freedom to participating writers. Each writer was responsible for creating and writing a patient visiting a psychiatrist. This literary device allowed the authors to write about anything they wanted. But this freedom about choice of subject matter put a constraint on the age limit for both readers and writers due to the adult themes that were discussed by the fictional characters with their psychiatrist. And so the site could only cater to an adult education audience.

The Company Therapist's basic story line was about the world of a fictional psychiatrist, Doctor Charles Balis. The Doctor's primary job was to treat patients who were employees of a large computer company in San Francisco. From the web demographics, we believed that this setting was the most familiar to both readers and writers of The Company Therapist. Through transcripts of therapy sessions, detailed doctor's notes, doodles, personnel files, correspondence, and a variety of other collateral materials, the reader could delve deeply into the lives of these fictional patients.

The Company Therapist project started in June of 1996 and ran through March of 1999. As an entertainment, The Company Therapist project created a fictional world with many characters, events, and story lines coexisting and evolving in time to give an illusion of reality to its readers. The result was vast literary landscape consisting of thousands of book-sized pages. To read more about this project, visit it on the Web.

Additional Thoughts and Further Readings

The one real goal of education is to leave a person asking questions.

—Max Beerhohm

You've reached the end of the book (almost). Congratulations. I know you've been thinking hard about all of the conceptual, interaction, and interface design features that you see all around you. I hope you find them easy to recognize them now.

Alan Cooper provides many detailed examples of how his company uses personas as part of their interaction design services in his book "The Inmates are Running the Asylum: Why High-Tech Products Drive Us Crazy and How to Restore the Sanity."

For an exciting account of bad door designs and other frustrating interaction designs, read Donald Norman's book "The Design of Everyday Things."

For a wonderful description on painful transitions between cultures, consider reading "Lost in Translation: A

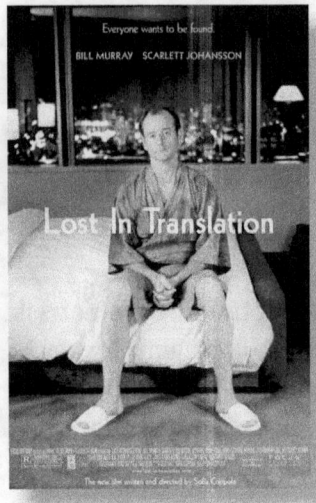

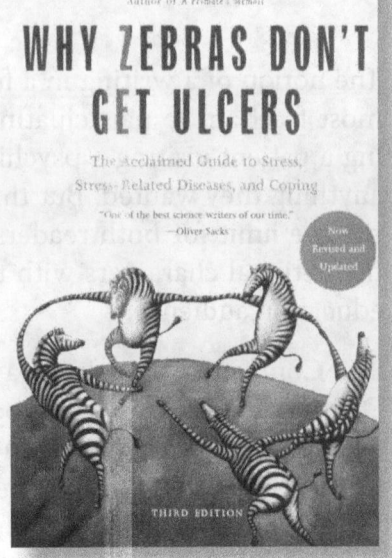

Life in a New Language" by Eva Hoffman. I particularly identify with Eva's story, who emigrated from Poland to Canada as a young teenager. I left Russia for America when I was thirteen—that's a difficult age to make cultural and language adjustments.

The 2003 movie by the same name, "Lost in Translation" by Sofia Coppola, is another interesting look at culture shock.

If you would like to learn more about conceptual, interaction, and interface design of comic books, consider spending some time with Scott McCloud's "Understanding Comics: The Invisible Art." It will be time well spent and you might even want to read some of his other wonderful books.

Robert Sapolsky spent his formative years studying baboons in Africa. His observations on culture and its influence on physiology are brilliant and funny. You might want to start with "Monkeyluv," a collection of previously published articles, and work your way to "A Primate's Memoir: A Neuroscientist's Unconventional Life Among the Baboons," a series of chronological essays about Dr. Sapolsky's research in Africa. I don't expect you to start darting tourists in museums to take their blood and analyze their stress hormone levels, but it's a thought. And if you like these, consider "Why Zebras Don't Get Ulcers: An Updated Guide to Stress, Stress-Related Diseases and Coping."

"More than Cool Reason: A Field Guide to Poetic Metaphor," written by George Lakoff and Mark Turner is an eye opener for those interested in metaphors. It's an interesting blend of language use and mental models.

By now you know that I'm a big fan of Oliver Sacks. And so I'm recommending another one of his books: "Uncle Tungsten: Memories of A Chemical Boyhood." This is a memoir of Dr. Sacks childhood in war-torn London in the 1940's. It's only been a few decades—a single lifetime—and yet the world and our perspective have shifted dramatically. For those of you interested in product design, you might want to read this book from the point of view of anthropology: What changed?

Finally, David Gelernter's visionary, although dated, book, "Mirror Worlds: or the Day Software Puts the Universe in a Shoebox...How It Will Happen and What It Will Mean," is a must for science fiction fans as well as product interaction designers. Dr. Gelernter thinks big and comes up with a futuristic model of computers embedded in the very fabric of society. The book was written in 1993, and he has published a few books since then, but for scope of technological futurisms, this one is my favorite.

Section Five: Additional Information

Bibliography

Appendix

My Background

20. Bibliography

I never let schooling interfere with my education.

—Mark Twain

Abrahamson, J., Greber, A. (2005). Manners Are Magic: "You'll Thank Me for Telling You" Lessons on Life from Ms. Manners. Emmis Books; ISBN: 1578602319

Barbour, R., Kitzinger, J. (1999). Developing Focus Group Research: Politics, Theory and Practice. Sage Publications Ltd; ISBN: 0761955682

Berners-Lee, T. (1999). Weaving the Web. New York, New York: HarperCollins Publishers.

Bransford, J. D., et al. (2000). How People Learn: Brain, Mind, Experience, and School. Washington, D.C.: National Academy Press.

Brown, A., Bransford, J., Ferrara, R., Campione, J. (1983). "Learning, Remembering, and Understanding". In J. Flavell & E. Markman (Eds.), *Handbook of Child Psychology* (4th ed.). Cognitive Development (Vol. 3). Section on Metacognition, pages 106-126. New York: Wiley.

Bryson, B. (1991). The Mother Tongue. Harper Perennial; ISBN: 0380715430

Carroll, J., Thomas, J. C. (1982). "Metaphor and the Cognitive Representation of Computing Systems." IEEE Transactions on Systems, Man, & Cybernetics. 12(2), 107-116.

Cassidy, D. (2007). How the Irish Invented Slang. CounterPunch Books and AK Press; ISBN: 1904859607

Cooper, A. (1999). The Inmates are Running the Asylum: Why High-Tech Products Drive Us Crazy and How to Restore the Sanity. Indianapolis, Indiana: SAMS, a Division of Macmillan Computer Publishing.

Csikszentmihalyi, M. (1991). Flow: The Psychology of Optimal Experience. Harper Perennial; ISBN: 0060920432

diSessa, A. A. (1983). "Phenomenology and the Evolution of Intuition." In D. Genter & A. L. Stevens (Eds.), *Mental Models*, pp. 15-33. Hillsdale, NJ: Lawrence Erlbaum Associates.

Dillenbourg, P. (1999). "What do you mean by collaborative learning?" In P. Dillenbourg (Ed.), *Collaborative-learning: Cognitive and Computational Approaches*. Oxford: Elsevier.

Druin, A. (1999). The Design of Children's Technology. San Francisco, CA: Morgan Kaufmann Publishers, Inc.

Edmunds, H. (2000). Focus Group Research Handbook. McGraw-Hill; ISBN: 0658002481

Fisher, S. (2001). Teaching and Technology: Promising Directions for Research on Online Learning and Distance Education in the Selective Institutions. Report for the Andrew W. Mellon Foundation. (Draft)

Gelernter, D. (1993). Mirror Worlds: or the Day Software Puts the Universe in a Shoebox...How It Will Happen and What It Will Mean. Oxford University Press, US; ISBN: 019507906X

Gregorc, A. F., Butler, K. (1984). Learning is a Matter of Style. VocEd, 27-29.

Halasz, F., Morgan, T. P. (1982). "Analogy Considered Harmful." Proceedings of the ACM Conference on Human Factors in Computer Systems, Gaithersburg, MD, 383-386.

Hoffman, E. (1990). Lost in Translation: A Life in a New Language. Penguin Books; ISBN: 0140127739

Jonassen, D. H. (1992). "Designing Hypertext for Learning." In Scanlon, E. & O'Shea, T. (Eds.) *New Directions in Educational Technology*. Springer-Verlag. Berlin.

Jonassen, D. H., Peck, K. L., Wilson, B. G., Pfeiffer, W. S., (1998). Learning with Technology: A Constructivist Perspective. Prentice Hall; ISBN: 013271891X

Jonassen, D. H., Howland, J., Moor, J., Marra, R. M., (2002). Learning to Solve Problems with Technology: A Constructivist Perspective (2nd Edition). Prentice Hall; ISBN: 0130484032

Jonassen, D. H., Grabowski, B. L. (1993). Handbook of Individual Differences, Learning & Instruction. Hillsdale, New Jersey: Lawrence Erlbaum Associates.

Kommers, P. A. M., Grabinger, R. S., Dunlap, J. C. (Editors). (1996). Hypermedia Learning Environments: Instructional Design and Integration. Lawrence Erlbaum Associates; ISBN: 0805818294

Keirsey, D., Bates, M. (1984). Please Understand Me: Character & Temperament Types. Del Mar, California: Prometheus Nemesis Book Company.

Kieras, D. E., Bovair, S. (1984). The Role of a Mental Model in Learning to Operate a Device, Cognitive Science, 8(3), 255-273.

Keirsey, D. (1998). Please Understand Me II: Temperament, Character, Intelligence. Del Mar, California: Prometheus Nemesis Book Company.

Keirsey, D., Bates, M. (1984). Please Understand Me: Character & Temperament Types. Del Mar, California: Prometheus Nemesis Book Company.

Kolb, D. A. (1984). Experiential learning: Experience as the Source of Learning and Development. New Jersey: Prentice-Hall.

Lakoff, G., Turner, M. (1989). More than Cool Reason: A Field Guide to Poetic Metaphor. University Of Chicago Press; ISBN: 0226468127

Lave, J., Wenger, E. (1991). Situated Learning: Legitimate Peripheral Participation. Cambridge: Cambridge University Press.

Levine, M. (2004). The Myth of Laziness. New York, New York: Simon & Schuster.

Levine, M. (2002). A Mind at a Time. New York, New York: Simon & Schuster.

Madel, T. (1997). Elements of User Interface Design. John Wiley & Sons, Inc.

McCloud, S. (1994). Understanding Comics: The Invisible Art. Harper Paperbacks; ISBN: 006097625X

Miller, G. A. (1956). "The Magical Number Seven, Plus or Minus Two: Some Limits on Our Capacity for Processing Information." Psychological Science, 63, 81-97.

Nielsen, J. (2000). Designing Web Usability: The Practice of Simplicity. New Riders Publishing; 1st edition. ISBN: 156205810X

Norman, D. A. (1988). The Psychology of Everyday Things. Basic Books. ISBN: 0465067093

Norman, D. A. (1983). "Some Observations on Mental Models." In Gentner, D. & Stevens, A. (Eds.), *Mental Models*. Mahwah, NJ: Lawrence Erlbaum Associates. (pp. 7-14).

Nunberg, G. (2004). Going Nucular: Language, Politics, and Culture in Confrontational Times. Public Affairs; ASIN: B000Q6SKP0

Papert, S. (1980). Mindstorms: Children, Computers, and Powerful Ideas. USA: Basic Books, Inc.

Sacks, O. (2002). Uncle Tungsten: Memories of a Chemical Boyhood. New York, New York: Vintage Books. ISBN: 0375704043

Sacks, O. (1998). A Leg to Stand On. Touchstone; ISNB: 0684853957

Sacks, O. (1998). The Man Who Mistook His Wife For A Hat: And Other Clinical Tales. Touchstone; ISBN: 0684853949

Sacks, O. (1996). The Island of the Colorblind. New York, New York: Vintage Books.

Sacks, O. (1990). Seeing Voices: A Journey into the World of the Deaf. New York, New York: HarperCollins Publishers Inc.

Safire, W. (2008). Safire's Political Dictionary. Oxford University Press, USA; ISBN: 0195340612

Sapolsky, R. M. (2005). Monkeyluv: And Other Essays on Our Lives as Animals. Scribner; ISBN: 0743260155

Sapolsky, R. M. (2002). A Primate's Memoir: A Neuroscientist's Unconventional Life Among the Baboons. Scribner; ISBN: 0743202414

Sapolsky, R. M. (1993). Why Zebras Don't Get Ulcers: An Updated Guide to Stress, Stress-Related Diseases and Coping. W. H. Freeman & Company; ASIN: B000WL06QU

Schank, R. C. (1990). Tell Me a Story: A New Look at Real and Artificial Memory. New York: Macmillan Publishing Company.

Shneiderman, B. (1998). Designing the User Interface: Strategies for Effective Human-Computer Interaction. Menlo Park, CA: Addison-Wesley Longman, Inc.

Simon, J. S., (1999). Wilder Nonprofit Field Guide to Conducting Successful Focus Groups. Fieldstone Alliance; ISBN: 0940069199

Smith, D. M., (1968). A History of Sicily, 800-1713: Medieval Sicily. Chatto & Windus.

Tannen, D. (2006). You're Wearing THAT?: Understanding Mothers and Daughters in Conversation. Random House, Inc. New York.

Tannen, D. (2001). You Just Don't Understand. Harper Paperbacks. ISBN: 0060959622

Tannen, D. (1986). That's Not What I Meant! Ballantine Books, New York, NY.

Underwood, J. D. M. & Underwood, G. (1990). Computers and Learning. Cambridge, MA: Basil Blackwell, Inc.

Vygotsky, L. S., (1978). Mind in Society, The Development of Higher Psychological Processes. Cambridge, Massachusetts: Harvard University Press.

Werby, O. (2008). "Visual Symbolic Processing in Modern Times," Ed-Media 2008, AACE, Vienna, Austria.

Werby, O. (2007). "Examination of Student Motivation and Group Dynamics in the Internet-based Learning Experiences," Ed-Media 2007, AACE, Vancouver, British Columbia.

Werby, O. (2007). "The Situational Learning Matrix: a Design Tool for Creation of Internet-based Learning Experiences," Ed-Media 2007, AACE, Vancouver, British Columbia.

Werby, O. (2005). "Development of Internet-based Learning Experiences; The Company Therapist Project," Doctorate Dissertation, University of California at Berkeley.

Werby, O. (1994). "The Relationship Between Changes in Perceptual Focus and Understanding," University of California at Berkeley, 1994 AERA Presentation.

Werby, O. (1993). "What Kids Know About Research," Master's Thesis, University of California at Berkeley.

Woolfolk, A. E. (1998). Educational Psychology, Seventh Edition. Allyn & Bacon, Needham Heights, MA.

Young, R. M. (1983). "Surrogates and Mappings: Two Kinds of Conceptual Models for Interactive Devices. "In Gentner, D. & Stevens, A. (Eds.), *Metal Models.* Mahwah, NJ: Lawrence Erlbaum Associates. (pp. 35-52).

There are 6 Fs.

There are 15 insects.

21. Appendix

Only the shallow know themselves.

—Oscar Wilde

Personality and Learning Styles Literature Background

Product design borrows from many different disciplines: psychology, education, marketing, physiology, just to name a few. Each of these disciplines has practitioners and researchers. There is a lot of new terminology thrown around, some of which catches on and the individuals that came up with it get recognition. There is a strong incentive to come up with new ways of expressing old ideas. Sometimes these new formalizations are helpful in understanding and solving problems within the discipline and sometimes not. Having multiple ways of expressing the same idea creates confusion and limits usefulness of existing data.

In particular, the concepts of personality and learning styles have been used, abused, and confused. To provide a bit of clarity, I've conducted a review of how the ideas and definitions of personality and learning styles have been used. Basically, when you hear "learning style" you should think "personality." Below is a quick summary of history and ideas in this field.

What Type Am I?: The Myers-Brigg Personality Type Indicators

Isabell Myers and her mother, Kathrin Briggs, developed a taxonomy of personality types based on the psychological teachings of Carl Gustav Jung. There are four scales (or dimensions): **Extroversion/Introversion**, **Thinking/Feeling**, **Judging/Perceiving**, and **Intuitive/Sensing**. Each extreme of a personality pair plots the opposite end of that continuum, and no person, according to this theory, can be at both ends at once. A particular individual's personality types is defined by the four variables out of the possible eight. By answering a set of questions, anyone can find their personality type and read the explanations which accompany them. There are numerous sites on the Internet that provide Myers-Briggs personality tests and information. The standard **Myers-Briggs Personality Indicator** test has been developed over thirty-five years ago and is widely used. Quick definitions for each of the four Myers-Briggs dimensions and their two spectrums is given below.

Introverted/Extroverted (Reserved/Expressive): Introversion implies that a person prefers to direct attention toward an inner world of concepts and ideas. Extroversion, in contrast, implies that a person prefers to direct attention toward external stimuli.

Intuitive/Sensing (Introspective/Observant): Intuitive implies that a person prefers to perceive the world through impressions and to imagine possibilities, rather than focus on reality. Jung used the word intuition to literally mean "internal attention." By contrast, sensing implies that a person prefers to take in the world through their senses, as it's presenting itself, to concentrate on the practical and the immediate.

Judging/Perceiving (Scheduling/Probing): Judging implies that a person prefers to make decisions, to come to conclusions. Perceiving implies that a person prefers to probe for options and not to be tied down by plans and schedules.

Thinking/Feeling (Rational/Emotional): Thinking implies that a person prefers to follow reason and thought. Feeling implies that a person prefers to concentrate on feelings—to follow one's heart.

You should note that all of the personality-based classifications are based on personal introspection of each individual—this is a self-assessed classification.

Please Understand Me: Keirsey's Personality Dimensions

Myers and Briggs published their book, "The Myers-Briggs Type Indicator," in 1962. The personality test they developed was used by Educational Testing Services for many years to conduct research in collages and high schools around the country. David Keirsey first came across this test in such a setting. He later adapted and manipulated the four scales and modified the test. His latest research (which includes the data on online users' personality distributions) can be found at http://keirsey.com.

David Keirsey's personality dimensions are based on the popular Myers-Briggs Personality Taxonomy. He uses the four personality dimensions as introduced by Myers (Extroversion and Introversion, Intuition and Sensing, Thinking and Feeling, Judgement and Perception), but provides alternative definitions and even alternative labels (**Expressive** and **Reserved**, **Observant** and **Introspective**, **Thinking** and **Feeling**, **Schedulers** and **Probers**).

While very popular, the Myers-Briggs definitions have flaws as pointed out by David Keirsey. A definitional error has crept into the test as the result of Jung's early use of vocabulary. In particular, extroversion, as defined by Myers, has more to do with observation than the popularized meaning of being socially at ease and verbally expressive. Similarly, introversion, as defined by Myers, has more to do with introspection than the widely used notions of being shy or socially aloof and reserved.

The Cognitive Wheel (see Chapter 5) was developed to construct a set of dimensions requiring, for the most part, self-assessment. I was concerned that confusions over popular terms, as opposed to the meanings provided by Myers, would introduce errors into any system based on these qualities. To reduce such errors, introverted and extroverted dimensions have been replaced by the reserved and expressive personality types for the Cognitive Wheel.

Introverted and extroverted confusion is not the only problem with the Myers-Briggs test. Judging and perceiving traits are really used to describe personal temperaments in relation to time allocation. In particular, **judging** individuals are said to live by the clock and to judiciously assign their time schedules. **Perceiving** people don't like to make plans and tend to live in the moment. We all have friends who are always late and have "trouble with time." They probably have a perceiving personality. On the other side, some of us know what we will be doing a month from now and budget our time accordingly—that's an example of a judging personality. Since the meanings of "judging" and "perceiving" as defined by Myers are not tightly related to the defining traits, the Cognitive Wheel adapts Keirsey's variant: "schedulers" and "probers."

The final difference between Myers' and Keirsey's personality identification is their treatment of intuition and sensing personality characteristics. Since "intuition" in Jung's definition means "listening to the inner voice" or "heeding the prompting from within," it refers to people who prefer to pay attention to their feelings and thoughts. Thus "intuition" can properly be summarized as "introspection" and "internalization." "Sensing," by contrast, refers to "observation" and "externalization." It is much easier to understand what it means to be "observant" as opposed to what it means to be "sensing." Analogously, "intuition" is a harder concept to relate to than "introspection." Thus "observant" and "introspective" were chosen as personality dimensions for the Cognitive Wheel.

While there are 16 possible personality subtypes, Keirsey specifies four basic personalities: "SP's," "SJ's," "NF's," and "NT's". These personality subtypes show the widest and clearest differences in temperament. The full personality subtypes, as described by the four letter combinations, are refinements on the basic four. Keirsey gives these statistical numbers for the 16 subtypes in a US population:

INTP (Architect), about 1% of population	ESFJ (Seller), about 13%
ENTP (Inventor), about 5% of population	ISFJ (Conservator), about 6%
INTJ (Scientist), about 1% of population	ESFP (Entertainer), about 13%
ENTJ (Field Marshal), about 5% of population	ISFP (Artist), about 5%
INFP (Questor), about 1% of population	ESTJ (Administrator), about 13%
ENFP (Journalist), about 5% of population	ISTJ (Trustee), about 6%
INFJ (Author), about 1% of population	ESTP (Promoter), about 13%
ENFJ (Pedagogue), about 5% of population	ISTP (Artisan), about 5%

Keirsey included "personality portraits" as exemplars of individuals who might fall into the pre-defined categories. His explanations and descriptions of attributes for each personality dimensions seem to have clear implications to product design.

Kolb Learning Styles

> *The secret of education is respecting the pupil.*
>
> —Ralph Waldo Emerson

In 1984, David Kolb proposed that the way people learn depends on how they perceive the world around them and on how they process that information. Kolb proposed a learning plane defined by two dimensions: the perceiving dimension forming a "feeling/thinking" axis and the processing dimension forming a "doing/observing" axis. An individual can be assigned particular values on the "feeling/thinking" axis and on the "doing/observing" axis where they are most comfortable and perform at their best. A brief explanation of **Kolb's Learning Styles** is presented below.

Different people take different approaches to an experience. Some immerse themselves in it, perceive it through all their senses, and then try to connect the experience to meaning. These people like to actually hear, see, and touch things. Then there are people who would prefer to stand back and analyze, separating themselves from the experience and trying to think through what's going on. These two examples are the extremes—most people do a little analyzing and a little immersing, although perhaps not at the same time. For each individual interacting with a particular experience, there is a place on this continuum where they feel most comfortable. Kolb described this learning dimension as the "feeling/thinking" axis or the "concrete/abstract" axis (this is the "collect" experience axis).

Kolb defined the next part of the learning process by how people process the information they've just experienced. Some people prefer to try things out right away. These are the "doers." Others prefer to think things through first. They integrate what they've just experienced with prior knowledge in an attempt to gain a better understanding of the whole before attempting the experience themselves. These people are the "observers." Again, these are the extremes. Most people fall somewhere in between. Kolb described this learning dimension as the "doing/observing" axis or the "active/reflective" axis (this is the "process" experience axis).

Kolb then went on to define specific learner types by their position in his learning plane. **Type One Learners** are the people that fall in the "concrete (feeling)" and "reflective (observing)" quadrant of the learning plane. They take-in information with their senses and then reflect on their experience and generalize. People who feel most comfortable learning this way are classified as imaginative. Another term for Type One Learners is **Accommodative**.

Type Two Learners are the people that fall in the "abstract (thinking)" and "reflective (observing)" quadrant of the learning plane. They perceive information abstractly and then

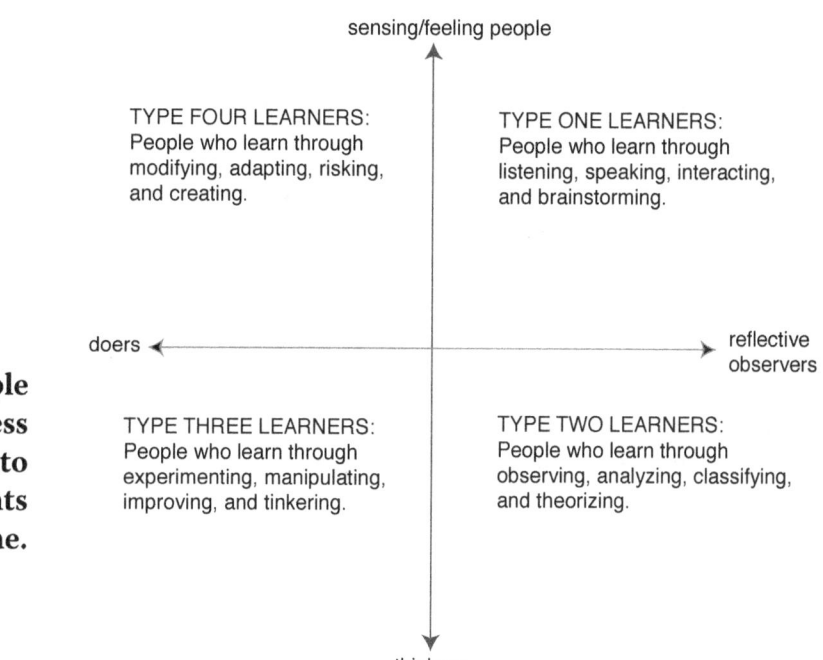

sensing/feeling people

TYPE FOUR LEARNERS:
People who learn through
modifying, adapting, risking,
and creating.

TYPE ONE LEARNERS:
People who learn through
listening, speaking, interacting,
and brainstorming.

doers ◄—————————————————► reflective
observers

TYPE THREE LEARNERS:
People who learn through
experimenting, manipulating,
improving, and tinkering.

TYPE TWO LEARNERS:
People who learn through
observing, analyzing, classifying,
and theorizing.

thinkers

Depending on how people perceive and process information, they fall into one of the four quadrants of Kolb's learning plane.

reflect on it. These are what people would call theoretical, analytical individuals. Another term for Type Two Learners is **Assimilative**. It's interesting to note that, in our society, most classroom instruction tends to be geared for Type Two Learners. In schools, people are encouraged to closely watch the lessons and then think through what they have observed. Classroom learning is mostly geared towards the memorization of facts and rules. But Type Two Learners make up only a fraction of the population.

Type Three Learners are the people that fall in the "abstract (thinking)" and "active (doing)" quadrant of the learning plane. These individuals think things through and then act on it, experimenting and testing things out. These are the practical, common sense people. Another term for Type Three Learners is **Converging**.

Type Four Learners are the people that fall in the "concrete (feeling)" and "active (doing)" quadrant of the learning plane. They feel their way through an experience and then process that information through action. These are the dynamic, intuitive types. Another term for Type Four Learners is **Diverging**.

By Kolb's definition, there are two learning axes: active/reflective and concrete/abstract. Kolb's Concrete individuals most closely resembles Myers-Briggs' Sensing Personality in meaning and implications. Both Sensing and Concrete dimensions identify individuals that prefer to use their senses to gather information and then to interact with it. Since I'm using Keirsey's set of dimensions for the Cognitive Wheel, Kolb's Concrete category is equivalent (or partially equivalent and collapsed into) Keirsey's Observant class of individuals.

Kolb's Abstract category most closely resembles Myers-Briggs' Intuitive set in meaning and implications. Both Abstract and Intuitive variables identify individuals that prefer the world of ideas and direct their attention towards thoughts rather than sensual perceptions. Again, using Keirsey's classification system, Kolb's Abstract learners are aligned with Keirsey's Introspective individuals.

Myers-Briggs' Introversion refers to people who direct their attention inwards to concepts and ideas. Extroversion refers to people who prefer to direct attention toward the external stimuli. As such, this Myers-Briggs' dimension most closely resemble Reflective/Active learning axis as defined by Kolb. While Introversion/Extroversion is not as closely related to Reflective/Active as Sensing/Intuitive to Concrete/Abstract, there is enough overlap that creating separate learner dimensions for both of these ideas will complicate the Cognitive Wheel system in the long run. And so finally, Keirsey's taxonomy which incorporates most of the Kolb's and Myers-Briggs' concepts is chosen for the Cognitive Wheel.

Gregorc Learning Styles

People learn more quickly by doing something or seeing something done.
—Gilbert Highet

A. F. Gregorc developed learning styles based on two dimensions: first, the individual's preference for perceiving information and, second, the individual's preference for organizing that information. Gregorc identifies four types of learners and differentiates among them using twelve variables. There is a **Concrete/Abstract** dimension and a **Random/Sequential** one. Gregorc dimensions are very similar in form and function to Kolb's Learning Styles; they even share some naming conventions.

Gregorc's **Concrete perception** refers to the individual's ability to process the physical aspects of information through the senses. These individuals prefer to gather information through hands-on experiences and have highly developed senses. Concrete is almost equivalent to Myers-Briggs' Sensing concept and Keirsey's Observant category. It is subsumed into the Cognitive Wheel's Observant Learner section.

Gregorc's **Abstract learning style** is the opposite end of the spectrum from Concrete. Abstract perception refers to the ability to process information through reason and intuition, as opposed to physical senses. Abstract individuals are characterized by their preference for idea and concept manipulation rather than experiential exploration. Again, Gregorc's Abstract learning styles is parallel to Kolb's Abstract learning style.

Gregorc's **Sequential learners** are defined by their need for order and logic. They need to organize information in order to understand it. To some degree, this category is very similar to the Myers-Briggs' Judging and Keirsey's Schedulers class of learners.

Gregorc's **Random learners** employ trial and error in the learning process. They have the need to experiment before coming to a conclusion and perform best in an unstructured

Sequential: need to organize information,
equal to Judging and Scheduling Personality.

Concrete = Sensing ——————————|—————————— Abstract = Intuitive

Random: need to experiment,
equal to Perceiving and Probing Personality.

setting. Some of the attributes of Gregorc's Random learners are parallel to Myers-Briggs' Perceiving and Keirsey's Probing dimensions.

Gregorc defined a particular learner by placing them somewhere on the Concrete/Abstract and Sequential/Random axes. Along with the four main distinctions made between learners, Gregorc learning styles provide a list of research-based instructional prescriptions that both support the individual's weaknesses and work to their strengths.

The Cognitive Wheel

The Cognitive Wheel is a tool for product designers and gathers the smallest set of orthogonal variables (don't you wish you still remembered math) from existing personality research literature. So here is how all of this research collapses into simplest set of variables.

Cognitive Wheel:	Keirsey:	Myers-Briggs:	Kolb:	Gregorc:
Expressive	Expressive	Extroverted	Active (Doing)	
Reserved	Reserved	Introverted	Reflective (Observing)	
Observant	Observant	Sensing	Concrete (Feeling)	Concrete
Introspective	Introspective	Intuitive	Abstract (Thinking)	Abstract
Schedulers	Schedulers	Judging		Sequential
Probing	Probing	Perceiving		Random
Emotional	Emotional	Feeling		
Reasoning	Reasoning	Thinking		

Of course I couldn't leave personality descriptions to just these eight variables and a few of my own. Chapter11: "Personality" has descriptions, slimmed and focused, that I recommend that you use in your product design research.

Other Research Linked to Field Dependence

As mentioned earlier, Field Dependence and Field Independence is a large catch-all for cognitive "stuff." The same ideas are discussed under many different names, but the educational and design solutions—the cognitive scaffoldings—are similar. Below is a brief discussion of some of the educational research.

Honey and Mumford created a learning style classification closely related to Kolb's Learning Styles. Their classification divides learners into Activists, Reflectors, Pragmatists, and Theorists. By their definition, **Pragmatic Learners** prefer to learn through deductive reasoning and to focus on problems. These learners function better when there is a single correct answer or solution. As such, Pragmatic Learners fall into the definition of the Field Dependence learners defined above. The design of prescriptive instructional materials or products aimed at this audience will be the same for both.

Cognitive Simplicity and **Cognitive Complexity** is another bipolar cognitive characteristic that closely resembles Field Dependence and Field Independence. Cognitive Complexity and Simplicity is based on the notion that individuals perceive, categorize, and organize their world into progressively more specific chunks of information. The degree of specificity of these chunks and the fineness of the differentiation is one of the units into which this bipolar dimension is divided into—the more complex an individual's world view, the more Cognitive Complexity she possesses. The other unit of measure of Cognitive Complexity is the degree of flexibility with which an individual can cope with complex external stimuli in her environment. Thus a Cognitively Simple learner will tend to form a relatively simple world view with good versus bad and strong versus weak categories, while a Cognitively Complex individual will have more flexible, relativistic categories which allow ambiguities and conflict.

G. S. Klein introduced **Cognitive Flexibility** in 1954 as a measure of a learner's ability to ignore distracting information and to zero-in on relevant data. Cognitive Flexibility ranges from **Constricted** to **Flexible**. Given a choice, it is good to be Flexible and bad to be Constricted. Constricted individuals have trouble inhibiting irrelevant information, have problems with concentration, and have difficulties adjusting their points of view when provided with information that is contradictory to their beliefs. Flexible learners, on the other hand, tend to adjust more easily to new information, are resistant to background informational noise, and have superior concentration. Cognitive Flexibility, thus, is very closely related to Field Dependence and Field Independence.

Grasha-Riechmann Learning Styles consider the social attitudes of the learners, and thus closely resemble the different coping strategies of individuals with strong orientations to a particular personality type. Grasha-Riechmann Learning Styles are the **Participant/**

Avoidant bipolar dimension, the **Collaborative/Competitive** bipolar dimension, and the **Independent/Dependent** bipolar dimension. The Independent/Dependent bipolar dimension is defined by the learner's attributes at either pole: Independent versus Dependent style. Independent learners can think for themselves, can work independently, tend to learn the material, listen attentively, and are confident in themselves and their abilities. Dependent learners tend to require authority figures to tell them what to do, limit themselves to learning the minimum required material, and have little intellectual curiosity. It is easy to see that Dependent learners are similarly defined to Field Dependent individuals, and Independent learners are akin to Field Independent individuals. And thus this bipolar Grasha-Riechmann Learning Style dimension is can be treated with the same design strategies.

Impulsivity and its opposite **Reflectivity** are yet another set of cognitive traits, and they are used to define temperament. **Reflective** individuals are able to control their initial response to external stimuli until they have time to think things through and reflect on the situation. The most obvious outcome for impulsive learners is their tendency for making careless mistakes—think attention controls. It is much better to be a reflective learner than an impulsive one. Novice learners tend to be more impulsive than more experienced ones. Clearly, reflectivity can be taught as a study and performance technique. And again, design solutions for impulsivity would be similar to interventions for Field Dependent individuals.

There are many more cognitive characteristics described in the research literature that seem to cover the same individual traits, but the ones described above seem to be the most mentioned. And for all of the above (including Field Dependence) it might be best to consider the individual's attention controls and degree of familiarity with school—the more time a person spends in a scholarly setting, the more strategies she develops for absorbing, processing, and understanding new information. Thus the user's amount of formal education is a strong predictor for the user's ability to learn new information and feel confident doing it.

Interface and Content Prescriptions

It is a very sad thing that nowadays there is so little useless information.

—Oscar Wilde

Once you've identified your audience, you need to be able to specify the type of content and interface that would support its needs. Below is a list of possibilities to choose from.

Interface and Content Prescriptions Summary:

- Instructional Sequence
- Community Content Type
- Content Form Type
- Level of Interactivity

- Level of Participation in a Community
- Size (Duration) of Interaction
- Emotional Impact
- Organizational Schemes
- Control over content, navigation, interface, etc.

Form of Information Gathering:

- Questions/Multiple Choice
- Questions/Freeform
- Poll
- Comment
- Contribution (art or text or other content)

Presentation Sequence:

- Inductive—from theory to examples
- Deductive—from examples to theory

...

- Easy to difficult sequence of ideas
- Challenging with difficult ideas first

...

- Wrapped in a Narrative Sequence
- Exploration
- Presentation

Community Interaction:

- BBS free
- BBS with a moderator
- Chat
- Email
- Instant messaging

Content Form Type:

- Video
- Animation
- Photograph
- graphics
- graphs
- icons
- text
- audio
- games
- simulations (implies real world)
- maps
- e-commerce or auction type content

Level of Interactivity:

High:

- Games
- Simulations
- BBS
- Chat
- Email
- Blog (text, video, and audio)

Medium:

- Exploration
- Polls
- E-commerce or auction (this could be a high level of interactivity)

Low:

- Text
- Plain Graphics (map, photographs, illustrations, icons)
- Audio and Video, if easily viewed by user

Level of Participation in a Community:

High:

- Writing for BBS or a Blog
- email
- Chat
- Contributing Art or other self-developed pieces of content
- Game playing
- Revealing Personal Information

Medium:

- Lurking on BBS
- Lurking in Chat rooms
- Polls

Low:

- Read or View content available to all

Size (Duration) of Interaction:

- small chunks
- medium chunks
- large chunks

Emotional Impact:

- Sad or Happy
- Controversial
- Politically Correct/Incorrect
- Surprise
- Uncertainty
- Novelty
- Fear
- Anxiety
- Enjoyment
- Neutral

Organizational Schemes:

- Outlines for the whole content
- Divisions into areas or chapters by content type
- Overviews
- Flowcharts
- Summaries
- Hierarchical Organizers, Diagrams, Schemes, etc.
- Non Hierarchical Organizers, Diagrams, Schemes, etc. (e.g. content-based like The Company Therapist project, found at www.TheTherapist.com)
- Story Boards
- Free Exploration
- Surprise Navigation
- Limited or revealed-just-in-time navigation

Modeling of Expected (or Appropriate) Behavior Prior to Launching Content:

- Foreshadowing questions
- Relating content back to personal experience

Control over Content, Navigation, Information, Interface, etc.:

- High
- Medium
- Low

General Interface Rules for Corporate Training or Educational Software

You can pay people to teach, but you can't pay them to care.

—Marva Collins

Any teacher that can be replaced by a computer, deserves to be.

—David Thornburg

Whether you're asked to create training software for corporate clients or educational titles for schools and home audiences, there are several simple guidelines that can help improve the overall interaction design and enjoyment of the product.

They are:

1. Allow the learner to quickly use her newly acquired knowledge in a meaningful way. Early implementation of a technique or an idea will help the learner incorporate it into her world view and will help motivate her to learn more.

2. Try to reduce the passive presentation of material to a minimum. Most people learn better while engaged in an activity rather than through passive reading or listening.

3. Present information in such a way that more complex notions are built upon simpler, more basic concepts. It's never a good idea to overwhelm the learner with information that doesn't make any sense to her. The balance in the design is between allowing the user to freely move between topics and revealing new information as the user gets comfortable with the prerequisite topics.

4. Create an atmosphere in which the user feels comfortable making mistakes. Instead of being traumatic, errors should be used as opportunities for learning. Remember, this is the time to make mistakes—it's better to crash the simulated plane than the real one.

5. If your goal is to teach about gravity, build upon real world experiences which the learner already has. Create situations where the learner constructs gravitational notions based on common sense. Begin to introduce mathematical formulas while allowing the student to develop an intuition for the gravitational behavior of objects. Show how the mathematical formulas for gravitation describe real world problems and accord with the student's newly developed gravitational intuition. Only then are the abstract formulas meaningful by themselves. Another approach may have sufficed to allow the student to solve numerical problems, but with the approach above, the formulas themselves will be meaningful abstractions of understood behavior.

6. The learner should be able to build her own theories about the world and then try them out and see if they work. The more involvement in the topic and the greater the emotional and intellectual investment made by the learner in her own learning, the higher the retention of the material and the better the understanding. Give control to the learner and let her make the topic hers.

Please note that there is a fundamental difference between instructional software for kids and training software for professionals: the problems that kids are likely to face are well-defined with clear goals and with little risk involved. This is not true for adults. Upon completing professional training, adult learners are placed in real-world situations where problems are very ill-structured, unpredictable, and unique. The decisions that these people make might well impact the lives of others. These differences should be taken into account while designing educational or informational software or web sites. For a simulation aimed at adult professionals to be meaningful, it should simulate the real world complexity that the professional will actually face. Otherwise, its content will not be easily adapted to the needs of the professional acting in the real world.

Motivation and assessment are closely linked. If a student does poorly on an exam, that student may be motivated to study harder—this is an example of motivation based on teacher's assessment. A woman who doesn't fit into a size eight dress may be motivated to exercise more often or go on a diet—this is an example of motivation based on peer-pressure. A writer feels inferior (or less successful) to his contemporaries and is motivated to take a class—this is an example of motivation based on self-assessment of one's ultimate potential and current state of expertise. For an instructional structure to succeed, it needs to build-in motivational scaffolds as part of assessment.

Sample Help for Online Learning

Computers are magnificent tools for the realization of our dreams, but no machine can replace the human spark of spirit, compassion, love, and understanding.

—Louis Gerstner, CEO, IBM

For readers interested in developing online learning opportunities, here are samples of learning strategies that can be offered to the students. The type of learning strategy chosen—audio summaries, graphical organizers, outlines—should depend on the student audience taking an online class. In the best case scenario, the sample help shown below is targeted to users based on their pre-tests and ongoing performance with the learning materials.

Summary:

> After finishing this learning object, please write down a brief summary of this material.

Audio Summary:

> After finishing this learning object, use a tape recorder or other recording device and create an audio summary of the key ideas and concepts of this material. Play your tape to yourself on the way to work in a car or at other opportunities. This will help you to better understand and remember this material.

Reader's Summary:

To help yourself understand and remember the information better, please write down a short summary for this learning object.

Graphical Summary:

After finishing this learning object, use pencils and paper or a computer-based graphical application to create a visual summary of the key ideas and concepts of this material. This visualization can be in a form of an illustration, or a concept map, or a graphical outline, whichever is more helpful to you. Put your illustration up on a wall or some place where you can easily refer to it later. The process of creating this visualization and subsequent referrals to it will help you to better understand and remember this material.

Key Ideas and Concepts:

What are the main ideas or concepts presented in this learning object? Please write them down in a bulleted list format.

Graphic to Text Outline:

Using the graphical outline in the learning object, please create a text-based outline of the content. Use tabs or indents to indicate which are sub-topics or sub-headings. How does your outline fit into the overall organization of the entire course?

Outlining:

Use the headings of each section in this learning object (including the ones you might have created) and create your own outline of this material. Use tabs or indents to indicate which are sub-topics or sub-headings. How does your outline fit into the overall organization of the entire course? Refer to this outline for a quick and frequent visual review of this learning object.

Headings:

Please read each paragraph in this learning object carefully. For each paragraph, please write down a title or heading that best expresses the ideas in that paragraph.

Paraphrasing:

> Please read the information presented in this learning object several times, then paraphrase the main ideas and write them down below.

Example/Non-Example Generation:

> After finishing this learning object, please write down two examples and two counter examples that best illustrate the main ideas of this learning object. To share your examples with others taking this class, please copy and paste it on the bulletin board related to this class.

Example Generation:

> After finishing this learning object, please write down four examples that illustrate the main ideas of this learning object. To share your examples with others taking this class, please copy and paste it on the bulletin board related to this class.

Questioning:

> After finishing this learning object, please write down any questions that you might still have about this material. You will be able to refer to these questions later to see if further instruction has answered your questions.

Test Questions:

> After finishing this learning object, please write down a list of test questions that would adequately test someone else's understanding of this material. Can you answer your own test questions? To share your questions with others taking this class, please copy and past it on the bulletin board related to this class.

Vocabulary Words:

> After finishing this learning object, please write down a list of words that were new to you or that were defined in the learning object. Write down your own definitions of these words. If the words were not defined in the learning object, look them up in a dictionary and paraphrase their definition. This list will become part of your Glossary, and you will be able to access it at any time during your instruction.

Analogies:

After finishing this learning object, please write down an analogy that best relates the content of this learning object to your life, work, or interests. If you can't think of a direct analogy to your personal life, try to relate this material to other situations that you think are appropriate.

Graphical Outline:

After finishing this learning object, use pencils and paper or a computer-based graphical application to create a graphical outline of this material. For help with graphical outlines, please consult our Help Section or use the modeled example below.

Personal Example:

After finishing this learning object, please write down an example from your personal life that demonstrates a possible application of this material. To share your story with others taking this class, please copy and past it on the bulletin board related to this class.

Implications:

After finishing this learning object, please write down some implications that this material might have to your personal life. For example, will the information in this learning object change the way you approach your work? Or will it change your relationship with your coworkers?

Case Study:

After finishing this learning object, please write down a fictional case study that exemplifies some of the key ideas covered by this material. To share your story with others taking this class, please copy and paste it on the bulletin board related to this class.

Reading Out Loud:

Please read the material in this learning object several times out loud. This will help you to better understand and remember this material.

Visual Memory Test

In a few minutes a computer can make a mistake so great that it would have taken many men many months to equal it.

—Anonymous

This is a simple visual short term memory test referred in Chapter 6. Please time yourself, allowing no more that four seconds to view the illustration bellow. Then return back to the section "Short Term Memory Test" Chapter 6 and draw the illustration from memory. Please do not spend more than a few minutes making your drawing. After you're done, flip the pages back and forth and compare your drawing with the illustration below.

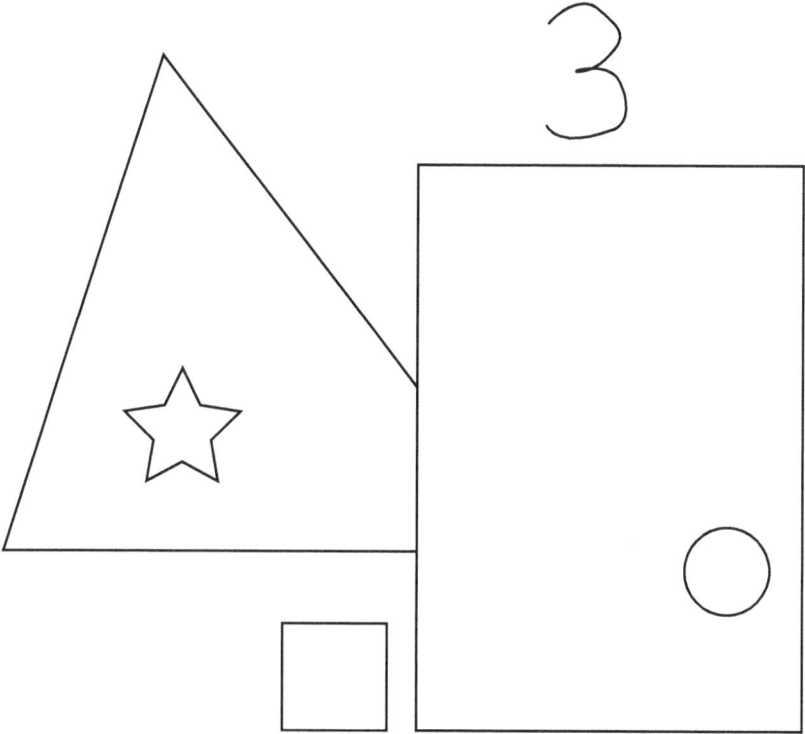

Did you include all of the elements? Are they in the right position? What are the differences between your drawing and this illustration?

About the Author

*A year spent in artificial intelligence is enough to
make one believe in God.*

—Alan J. Perlis

Olga Werby has a Doctorate from U.C. Berkeley's School of Education with a focus on online learning experience design and a Master's degree from U.C. Berkeley in Education of Math, Science, and Technology. She has been creating computer-based projects since 1981 with organizations such as NASA (where she worked on the Pioneer Venus project), Sunburst Communications, Addison-Wesley, and the Princeton Review. Olga has a Bachelor of Arts degree in Mathematics and Astrophysics from Columbia University. She was part of the faculty of San Francisco State University's Multimedia Studies Program, the Bay Area Video Coalition, and the campus of Apple Computers where she taught Interaction Design and Cognitive Theory. Olga holds a California teaching credential and often tests science-related curriculum materials in the San Francisco Unified School District public elementary and middle schools.

Olga, together with her husband Christopher Werby, formed Pipsqueak Productions, LLC in 1994. Pipsqueak is a boutique Web design and digital production firm. To learn more about Pipsqueak and what they do, please visit their Web site: www.Pipsqueak.com.

www.ingramcontent.com/pod-product-compliance
Lightning Source LLC
Chambersburg PA
CBHW081109170526
45165CB00008B/2387